ANTARCTICA: ENVIRONMENT AND CONSERVATION

(Arising Issues, Policies and Management)

Neloy Khare
Anoop Kumar Tiwari
Tara Megan Da Lima

2023

Published by

SCIENTIFIC PUBLISHERS (INDIA)
5 A, New Pali Road, P.O. Box 91
Jodhpur 342 001 (India)
E-mail: info@scientificpub.com
Website: www.scientificpubonline.com

ISBN : 978-93-92590-83-2

Printed in India

Dedicated to ...

Late Dr Milind P. Wakdikar
(28.05.1966 – 19.05.2021)

Dr Milind P. Wakdikar, after obtaining his B.Tech, M.Tech and PhD from IIT, Mumbai, he joined the Ministry of Earth Sciences (Erstwhile Department of Ocean Development). He dedicated his life to the furtherance of Ocean Sciences and Technology while serving the Government in various capacities including his long tenure as the Adviser to the Government of India at MoES. He rendered yeoman service to Earth Sciences for over 28 years. As a member of the Legal and Technical Commission, International Sea Bed Authority, United Nations, he has been the pillar of the Indian Ocean Mining Programme. He departed after a valiant fight with COVID-19. With his sudden demise on May 19, 2021, we lost a fine gentleman, a true friend, philosopher and affectionate colleague. Late Dr Wakdikar remained a man of principle and of a generous heart who always stood for the cause of the right things. All his colleagues at the Ministry of Earth Sciences, the Government of India and its various autonomous Institutes will always miss him. As a small tribute to Late Dr Wakdikar, we humbly dedicate this book to him.

Neloy Khare

Anoop Kumar Tiwari

Tara Megan Da Lima Leitao

Foreword

The frozen landmass larger than India and China, situated at the extreme southern latitude, is more than just spectacular but vital for our survival. Antarctica is a desert with not much rain. The snow does not melt when it snows and builds up over many years to make significant, thick sheets of ice called ice sheets. Antarctica is made up of lots of ice in the form of glaciers, ice shelves and icebergs.

The Antarctic ice deflects some of the sun's rays away from the earth's surface, thus maintaining its temperature. The study of Antarctica also allows us to understand the earth's present and help assess the scenarios. Gleaning clues from Geological history are trapped in Antarctica's various natural archives; we can reconstruct the earth's past. About 90 per cent of the earth's total ice is stored in Antarctica.

Antarctica remains almost totally covered by glacier ice; the Transantarctic Mountains extend across the continent, forming a convenient physical boundary to demarcate East Antarctica from West Antarctica (or Lesser Antarctica). Antarctica carries many superlatives and thus is unique among all the continents. These distinctive characteristics include being a desert and one of the most rapidly warming areas on earth, where there is no Antarctic time zone, but it is a place of active volcanoes and many subglacial lakes.

Antarctica's Ecosystem, including surrounding seas, is highly productive because phytoplankton grows abundantly during the extended daylight of summer and feeds huge krill populations. Krill, shrimplike crustaceans, are a crucial animal in this ecosystem, as food for top predators: whales, penguins, and seals.

Antarctica is connected to the rest of the world through oceanic and atmospheric circulations as critical drivers of the earth's oceanic and atmospheric systems. A critically important feature is that 70% of all available freshwater is locked up in the Antarctic ice sheet - if melted, this would raise sea levels by nearly 60 metres.

Equally important, between the coastline of the Antarctic continent and the Antarctic Polar Frontal Zone (the boundary between subtropical and subantarctic waters) lies the Southern

Ocean, which extends over some 38 million km2 and encompasses about 20% of the world's ocean waters.

Due to uncontrolled anthropogenic activities and global warming, the Antarctic also gets exposed to various threats like Climate change which is the greatest long-term threat to the region, increased fishing pressure and illegal fishing, and marine pollution. Persistent organic pollutants (POPs) and Invasive species. To conserve the pristine environment of the Antarctic, the Environmental Protocol of the Antarctic Treaty, sometimes called the "Madrid Protocol", became law in 1998 after legislation in each of the member countries.

The Act also prohibits any minerals-related activities, other than following a permit for scientific research; the harming or disturbing of the native flora and fauna; or the introduction of non-native species or entry into areas protected under either the Protocol or the Convention for the Conservation of Antarctic Marine Living Resources (CCAMLR), except per a permit. It is illegal under the Act as amended by the Antarctic Act 2013 to damage or destroy objects from a historical site or monument.

India has a long journey of scientific pursuits in the Antarctic region. Our scientists have demonstrated unparalleled insight and expertise in the new areas of Antarctic science and logistics. We have launched over 40 scientific expeditions to the icy continent, and it is essential to make an ordinary person aware of the Antarctic environment's danger. The present book is most aptly conceived and conceptualized to provide a ready guide detailing various dimensions of Antarctica's fragile environment to all travellers to the icy continent of Antarctica and to all vivid readers who are curious to know about the multi-facets of Antarctica.

The book **"Antarctica: Environment and Conservation (Arising Issues, Policies and Management)"** covers Antarctica's environmental issues and acts as an essential guide to anyone who intends to visit Antarctica. I compliment all authors for their commendable job in bringing out this timely book which will emerge as a reference for all, especially the young generation.

(Prem Chand Pandey)

Founder Director,

National Centre for Polar and Ocean Research,

Goa

Preface

Over a century people have travelled to many parts of Antarctica. Due to such frequent disturbances to the pristine environment of Antarctica, some species of Antarctic animals have been taken to the verge of extinction for economic benefit while others have been killed incidentally or disturbed, soils have been contaminated, untreated sewage has been discharged into the sea and rubbish has been left behind in even the remotest parts that will not decompose or break down for hundreds of years.

Fortunately, there has been a change in our attitude and we have begun to realise that there are few untouched and unvisited places left on earth and that they are tremendously important to humanity.

The ever-increasing demand to visit the Earth's last great wilderness tempted various tour operators to expand tourism-related activities over the Antarctic region which has added threat to Antarctica's clean air, water and ice which have been considered important to science for understanding how the Earth's environment is changing both naturally and as a result of human activity.

Global change may have effects that impact directly the Antarctic environment and its fauna and flora. Global warming may contribute to the break-up of ice-shelves causing loss of habitat for animals dependent on the iceshelf as well as the effect of increasing sea level on low-lying regions in the rest of the world. Increasing Ultra Violet (UV) radiation due to the ozone hole may cause changes to phytoplankton communities and thus the food chain.

The major threats Antarctica is being faced by Climate change/Global warming, Fishing, both legal and illegal, Invasive species, Tourism, Pollution, Exploration and exploitation of mineral reserves, oil and gas and Direct impacts associated with the development of infrastructure for scientific bases and programmes besides others.

It was therefore, imperative to evolve the policies, guidelines and legal framework to efficiently manage the Antarctic environment in a bid to restore past damage and how reduce the current and future impacts.

To assess the impact that the research base and its activities are having on the area around it, Environmental audits are now being carried out around scientific stations, on land and in the sea. With this intention, in the present book titled **"ANTARCTICA: ENVIRONMENT AND CONSERVATION"**, we have discussed the Arising Issues, Policies and Management related to Antarctic Environment in a nutshell. It is hoped that this book will serve as a ready reckon for all visitors to the icy continent of Antarctica.

We are hopeful that our humble efforts and contributions towards conserving Antarctic Environment will go a long way and will be a useful guideline for everyone who carries values and care for Antarctica.

Neloy Khare

Anoop Kumar Tiwari

Tara Megan Da Lima

Acknowledgements

The authors express their gratitude to the Secretary, Ministry of Earth Sciences and Director, National Centre for Polar and Ocean Research, Goa for their encouragement and kind support. The authors gratefully acknowledge the kind support and help received from many individuals working in the Polar regions. The materials/photos/images obtained from various websites are deeply acknowledged. All online contributors/ photographers and various researchers are sincerely acknowledged whose inputs are used as the background material for this book purely for academic purposes for disseminating the information among students and avid readers. Polar Scientists whose photographs are used in this book are deeply acknowledged. Publishers the Scientific Publishers, Jodhpur have done a commendable job and are sincerely acknowledged.

Neloy Khare

Anoop Kumar Tiwari

Tara Megan Da Lima

About the Author's

NELOY KHARE

Dr Neloy Khare, presently Adviser/Scientist 'G' to the Government of India at MoES has a very distinctive acumen not only in administration but also in quality science and research in his areas of expertise covering a large spectrum of geographically distinct locations like Antarctic, Arctic, Southern Ocean, Bay of Bengal, Arabian Sea, Indian Ocean etc. Dr Khare has almost 30 years of experience in the field of paleoclimate research using paleobiology Palaeontology)/ teaching/science management/administration/ coordination for scientific programmes (including Indian Polar Programme) etc. Having completed his doctorate (PhD) on tropical marine region and Doctor of Science (D.Sc) on Southern High latitude marine regions towards environmental/climatic implications using various proxies including foraminifera (micro-fossil), have made significant contributions in the field of Paleoclimatology of Southern high latitude regions (the Antarctic and the Southern Ocean) using Micropaleontology as a tool. These studies coupled with his paleoclimatic reconstructions from tropical regions helped understand causal linkages and teleconnections between the processes taking place in Southern high latitudes with that of climate variability occurring in tropical regions. Dr Khare has been conferred Honorary Professor and Adjunct Professor by many Indian Universities.

He has a very impressive list of publications to his credit (125 research articles in National and International Scientific journals; 3 Special Issues of National Scientific Journals as Guest Editor; Edited Special Issue of Polar Science (Elsevier), Journal of Asian Earth Science (Elsevier), Quaternary International (Elsevier) and Frontiers in Marine Science as its Managing / Guest Editor. Authored/ edited many books, 130 Abstracts have been contributed to various seminars; 23 Popular Science Articles; 5 Technical reports). The government of India and many professional bodies have bestowed him with many prestigious Awards for his humble scientific contributions to Past climate changes/ Oceanography / Polar Science and Southern Oceanography. The most coveted award is the Rajiv Gandhi National Award -2013 conferred by Honourable President of India Other include ISCA YOUNG SCIENTIST AWARD, BOYSCAST FELLOWSHIP, CIES FRENCH FELLOWSHIP, KRISHNAN GOLD MEDAL, BEST SCIENTIST AWARD, EMINENT SCIENTIST AWARD, ISCA Platinum Jubilee Lecture, IGU Fellowship, besides many. Dr Khare has made tremendous efforts to popularize ocean science and polar science across the country by way of delivering many Invited lectures, radio talks and publishing popular science articles.

Dr Khare sailed in the Arctic Ocean as a part of "Science PUB" in 2008 during the International Polar Year campaign for scientific exploration and became the first Indian to sail in the Arctic Ocean.

ANOOP KUMAR TIWARI

Dr Anoop Kumar Tiwari is working as Scientist 'E' in the National Centre for Polar and Ocean Research (NCPOR), Goa. Dr Tiwari has over 21 years of experience in the areas of Carrying Capacity Projects, Environmental Impact Assessment (EIA) studies for Cement, Thermal Power Plant, Steel, Fertilizer and Hydrocarbon industries, in EIA of Antarctic stations, Environmental Monitoring, and Aerosol Study etc. He is representing India in the Committee for Environmental Protection (CEP) in the Antarctica Treaty Consultative Meeting (ATCM) as well as Leader of the Environmental Expert Group (EEG) of the Council of Managers of the National Antarctic Programme (COMNAP). He pursued a Bachelor of Engineering in Civil Engineering followed by a Master of Engineering in Environmental Engineering and completed his PhD in Marine Science from Goa University. He participated in eight Indian Scientific Expeditions to Antarctica under an environmental impact assessment study. Dr Tiwari has worked on environmental quality at Maitri station, invasive alien species, bio-prospecting, the effect of glacier melting at Priyadarshini Lake, and assessment of mercury and persistent organic pollutants (POP) in the polar region.

He was awarded the Shrimati Shiromadevi Prize (2008) for the paper 'Water Abundance and Effect of Glacier Melting at Priyadarshini Lake in Antarctica', in the Environmental Engineering Journals of the Institutions of Engineers. He also won 'The Rekha Nandi and Bhupesh Nandi Prize' (2009) for the paper 'Waste water fate at Maitri in Antarctica, Journal of the institution of Engineers (Environmental Engineering). He is also a member of The Indian Science Congress Association, National Environmental Science Academy and Indian Aerosol Science and Technology Association. Dr Tiwari has published more than 15 research papers in various national and international journals.

TARA MEGAN Da LIMA

Ms Tara Megan Da Lima Leitao has a master's degree in Biotechnology from Goa University (2014-2016; 3rd rank holder) worked with renowned pharmaceuticals such as Cipla and Lupin in the field of Quality Control. She is currently working on the long-term Environmental Impact Assessment project of the Indian Antarctic stations.

Contents

1

INTRODUCTION

1.1 "Antarctica" Some Facts

Antarctica is the land of superlatives, with the lowest humidity and temperature accompanied by the highest wind velocity (Duarte *et al.*, 2018). It appears to possess absolute wilderness on the planet with diverse landscapes and climates that could thrill anyone in Antarctica (Ossenkamp, 2001).

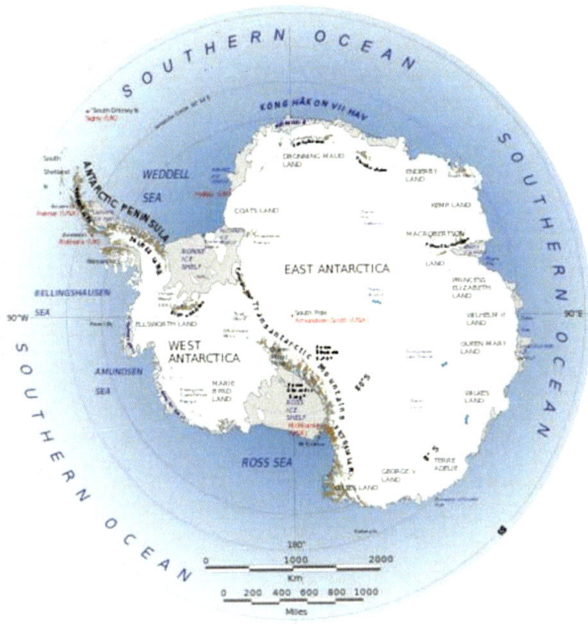

Figure 1.1: 'Antarctica' the icy Continent[1]

1.2 Natural Reserve

Antarctica plays an important role in the global climate circulation system and geodynamics. The cataclysmic separation of this continent from Gondwanaland millions of years ago and the following eras culminated in landmarks in geological history (Boger, 2011). According to the Antarctic Treaty System, anyone can use Antarctic land only for scientific experiments and peaceful purposes. Scientific minds have speculated Antarctica as an untapped source of fossil fuels embedded deep in the crust and benthic bed (Behrendt, 1983). It is a prime location for scientific assessments due to its environment's pristine and unsullied nature (Kennicutt *et al.*, 2014). Scientific research encompasses communication and satellite imaging, meteorological and atmospheric studies, environmental sciences, engineering, biology, physics, and medicine. Antarctic oceanography is not limited to the Southern Ocean but extends into the lower region of the Indian Ocean (Carter *et al.*, 2008).

1.3 Climate

1.3.1 Temperature

In Antarctica, the severity of temperatures depends (Figure 1.2) on the variation in latitude, elevation, and distance from the sea (Ma *et al.*, 2010). Due to higher elevation in East Antarctica and the polar vortex in the troposphere, the temperature is lower than in the Westward Antarctic continent (Van den Broeke and Van Lipzig, 2003). Comparatively, the Antarctic Peninsula projecting from West Antarctica into the Southern Ocean has a moderate climate showing a warming trend (Turner *et al.*, 2005). The Russian Station Vostok is considered to be one of the coldest points in the World (Turner *et al.*, 2005; Turner *et al.*, 2009a) It is located in the middle of the continent, far from the influences of the ocean. Using satellite remote sensing imagery Scambos *et al.* (2018) reported the lowest temperature on record as -98 °C on several occurrences (2004 - 2016), near the pole of inaccessibility.

Figure 1.2: Snowfall at Maitri[2]

1.3.2 Katabatic Winds and Rain in Icy Desert

Katabatic winds are driven by gravity and radially travel towards the ocean. These often translate into 'blizzards' formation by snow-endorsed cooling (Kodama *et al.*, 1985). Katabatic winds are known to travel at speeds 320 km/hr from the continent's centre toward the coasts (Brockett, 2005). Due to the barren expanse, it is considered a cold desert. Comparatively, Antarctica has about 50 mm of inland annual precipitation, less than the Sahara Desert. Annual precipitation along with the coast increases but only to about 200 mm equivalence of water. The interactions of cold and warm katabatic winds result in cyclonic storms leading to snowfall (Figure 1.3) along with the coastal areas (Carrasco *et al.*, 1997). Since the 1960s, the rate of snowfall precipitation has increased by 30%, less than an inch in depositions (Mosley-Thompson *et al.*, 1999).

Figure 1.3: High Wind with Freezing Temperature[3]

The significant difference between the Antarctic desert and other World deserts is the low rate of evaporation. Due to this, the surface snowfall remains unchanged thus forming ice sheets. Conversely, low evaporation rates allowed a degree of sub-surface ice evaporation (McKay *et al.*, 1998). During the 20th century, it is reported that the snow melts in the Peninsula have accelerated with time (Abram *et al.*, 2013). According to Fretwell *et al.* (2013), the build-up in Antarctic ice sheet volume has occurred gradually.

1.3.3 Climate Region

Antarctica can be divided into three superimposed rings based on climate. The outermost periphery has the highest precipitation rates, which decrease to the interior, which is severely cold with low precipitation rates. The coast has milder conditions, and the Peninsula region has the warmest and wettest climates.

1.4 Natural Resources

The minerals found in this continent are: iron ore, chromium, copper, gold, nickel, platinum, cryosols, and coal and hydrocarbons have been found in small non-commercial quantities (Rich, 1982; Behrendt, 1983; Riffenburgh, 2006; Simas *et al.*, 2006). The Antarctic marine biological resources (krill, crab, and fish) have been subjected to commercial fisheries. According to the Antarctica Treaty, exploration and exploitation of mineral resources are entirely banned, except for scientific purposes.

1.5 Neighbouring Continents

The continent's shape is an irregular circle extending to the Antarctic Peninsula into the Southern Ocean close to South America, depicted in (Figure 1.4). Coastal indentations include the Ross and Weddell ice shelves and their adjoining seas. The neighbouring continents are South America (1000 km), Australia (2500 km), and South Africa (4000 km).

Figure 1.4: Antarctica from Satellite[4]

1.6 Elevation

Charles Bentley was the man who discovered the lowest point in the Antarctic (Bentley, 2014). He has ventured on many treks and has also helped find many mountains. The Bentley sub-glacial trench is the lowest point (-2555-meter MSL) whereas, the highest point is Vinson Massif (4897-meter AMSL). The continent's elevation is 2.4 km above MSL which 1.5 km is more elevated than the global average.

1.7 Area

The icy continent consists of approximately 12.5 million km2 of which more than 99% comprises thick ice sheets, and approximately 0.5% is a barren rock (Burton-Johnson *et al.*, 2016). Some areas without ice cover include a part of southern Victoria Land (Bull, 1966; Beyer and Bolter, 2000), the Antarctic Peninsula, and Ross Island's stretch (Broady, 1989). A remarkable fact is that 11% of the continental area is contributed by floating ice shelves formed by glaciers that occupy half of the coast.

2

SCIENTIFIC IMPORTANCE

2.1 Significance of Research in Antarctica

Antarctica is an integral portion of the Earth's circulation system. By substituting as the heat sink of the World, it governs to control our climate and weather. Ice sheet security is a constant concern for those dwelling in areas of low elevation. If ice melting in Antarctica starts, it will raise the global sea levels to an estimated seventy meters from the present station. According to an estimate, the Antarctic continent has trapped 70% of the Earth's freshwater and 91% of the ice on Earth (Kennicutt *et al.*, 2014). Antarctica influences all the oceans; cold, dense, oxygen-rich waters originate in Antarctica and replenish the ocean's supply of bottom water, helping to drive ocean circulation (McKay *et al.*, 2012). The Antarctic sea supports all marine life sizes, from microscopic algae to giant whales (Knox, 2006). Discoveries and observations pave the way to a better understanding of how this environment responds to change.

2.2 Governance and Purpose

Antarctica is governed by the Antarctic Treaty System (ATS) (Tiwari, 2008). Antarctica Treaty was formed in 1959 by twelve nations and decided to use Antarctica land only for scientific experiments, investigations, and peaceful purposes (Tin *et al.*, 2008; Dodds, 2010). The Antarctic Treaty applies to the area located to the south of 60°S latitude, including all ice shelves. At present, 54 states are signatories to the Antarctic Treaty that includes 29 Consultative parties. India accessed the Antarctica

Treaty on August 19, 1983; it received consultative status on September 12, 1983. India ratified the "Protocol on Environmental Protection" to the Antarctica Treaty on January 14, 1998. The concept of setting aside areas for special protection of Antarctic areas (ASPAs) was introduced in 1964 when Consultative Parties decided to conserve endemic life on the continent (Convey *et al.*, 2012).

Due to the frontier science program and India's relentless commitment to the principles of the Antarctic Treaty, the following milestones were achieved:

- India signed the Antarctic Treaty and entered into force on August 19, 1983, and received consultative status on September 12, 1983;
- India's membership in the Scientific Committee on Antarctic Research (SCAR) was conferred on October 1, 1984.
- Convention on the Conservation of Antarctic Marine Living Resources (CCAMLR) was signed at Canberra on May 20, 1980, for the preservation and conservation of marine living resources in Antarctica. India ratified the said Convention on June 17, 1985 and is a member of the Commission for Conservation of Antarctic Marine Living Resources under that Convention (Pandey, 2005).
- India became a member of the Council of Managers of the National Antarctic Programme (COMNAP) in 1989.
- India also became the chairman of CCAMLR in 1998 for two years.
- The Protocol on the Environmental Protection to the Antarctic Treaty (Madrid Protocol) was signed at Madrid on October 4, 1991, to strengthen the Antarctic Treaty system and to protect the Antarctic environment and dependent and associated ecosystems. India signed the Madrid Protocol on January 14, 1998 (Tiwari, 2008).

2.2.1 Antarctica Treaty Consultative Meeting (ATCM)

Initially, the Antarctic Consultative Meeting (ATCM) was held every two years from 1961 to 1994; after that, it was held annually. This meeting upholds consultative parties (29 countries) and non-consultative parties (25 countries) (Chen, 2013). From 54 participatory countries, the consultative parties are the only decision-makers to all venturing to Antarctica.

2.2.2 Committee for Environmental Protection

The Committee on Environmental Protection (CEP) was formed under Articles 11 and 12 of the Antarctic Treaty and meets alongside the ATCM. It was established in 1998 and consisted of specialized groups. CEP has 41 members, out of which 29 are consultative parties, and the remaining (12) are non-consultative parties.

2.2.3 Council of Managers of National Antarctic Programmes (COMNAP)

COMNAP stemmed through ATCM, an international association formed in 1988 and is a platform for Antarctic Environmental and Logistics managers responsible for their National Antarctic Programmes in support of scientific research. (COMNAP, 2005) These managers work on behalf of respective Countries' Governments and in the spirit of the Antarctic Treaty. COMNAP has 30 National Antarctic Programme countries as a member.

2.2.4 Commission for Conservation of Antarctic Marine Living Resources (CCAMLR)

CCAMLR is an international multi-lingual conglomerate of 26 members (25 countries and the European Union) and ten acceding states. It was established in 1982, with its secretariat located in Hobart, Tasmania. CCAMLR is further responsible for multilateral collaboration to support the conservation of Antarctic marine life. To manage marine resources, CCAMLR employs the best scientific solutions and their applications.

2.3 India's Scientific Enticement to Antarctica

Antarctica provides a vast domain for insights into the 'magnetic field' of our planet. The interaction of charged particles from the sun and the promise of untold deposits of minerals and fossil fuels in the Antarctic landmass and the surrounding sea-bed are crucial in understanding magnetic moments. Scientists found Antarctica to be a valuable and essential place to research in various disciplines. The importance of Antarctica reflects in the following paragraphs:

- It is the only area in the southern hemisphere where simultaneous activities of the ionosphere and the planetary magnetic field can be observed (Sabaka *et al.*, 2018). The presence of natural solid electromagnetic signals forms an ideal environment for conducting studies on radio-wave propagation and radio noise in the ionosphere and the lower atmosphere (Simoes *et al.*, 2012).
- In the Mesozoic era, the super-continent 'Gondwanaland' was composed of five continents: Africa, Antarctica, Australia, Arabia, and South America (Segev, 2002). Over time the continents drifted apart, separated by the oceans (Tarling, 1971). Hence, Antarctica could divulge the secrets of world history.
- Geologists believed, Antarctica its surrounding sea-bed has the potential of unexploited natural reserves. As a result, Antarctica is a continent of great scientific value.
- The poles sustain the heat absorbed and emitted by the Earth. Atmospheric and oceanic heat dissipates as longwave radiation. Therefore, the Antarctic continent's cold air collides with the warmer air of the lower altitudes forming clouds (Turner *et al.*, 2009b). Accordingly, Antarctica regulates the global climate and particularly of the Southern hemisphere.
- The major oceans (Indian, Atlantic, and Pacific) join the Southern Ocean forming an uninterrupted ocean section.

The mixing process between cold and warm water demarcates Antarctic convergence's clear area with its unique physical, chemical, and biological characters (Talley, 2013).

- Antarctica provides a natural environment for conducting scientific observations. It is located a great distance from environmental pollution sources and constitutes an unpolluted datum point for monitoring global change (Bargagli, 2008).
- A landmass encloses the Indian Ocean at its northern end with its opposing end connected to the Southern Ocean. So it derives most of its productivity and nutrients from it (Gandhi *et al.*, 2012)
- The ocean of Antarctica plays an integral role in the oceanography of the Indian Ocean. Since parts of these oceans are connected, it is imperative to study the merged section in detail (Williams *et al.*, 2010).
- Antarctica holds the highest amount of freshwater reserves in a solid state (Shepherd *et al.*, 2018).
- The Antarctic Ocean supports rich benthic and pelagic communities (Gutt *et al.*, 2010). An important organism regulating the marine food chain in the Antarctic is "krill" (like shrimp), and its management is the pinnacle to sustaining marine ecosystems (Nicol and Foster, 2016).

2.4 Gateway to Antarctic Research

In the year 1981-1982, Indian scientists made their premier expedition to Antarctica which laid the foundation for scientific research. The focus of this voyage was to outline prospective research programs.

Situated in Goa, the National Centre for Antarctic and Ocean Research (NCAOR), an autonomous institute under the Ministry of Earth Sciences, Government of India, was the first institute in India to organize, coordinate and administer the Indian Scientific Program in Antarctica. It is currently known as the National Centre for Polar and Ocean Research NCPOR). Some scientific avenues include meteorology, geomagnetism, environmental science, radio-wave propagation, geology, glaciology, chemistry, microbiology, and lacustrine dynamics. The oceanographic studies consist of physical, chemical, biological, geological, and geophysical observations of the Southern Ocean and parts adjoining the Indian Ocean.

2.5 Dakshin Gangotri Station

The initiation of the Indian Scientific Expedition to Antarctica began in 1983 with the Dakshin Gangotri research station's development at the location (Latitude 703 04′ 59.99″ S Longitude 123 00′ E); nearly six km from the coast (Figure 2.1). It was an automated station powered by solar panels, housing Indian scientific instruments, and self-sustainable (Nair and Gupta, 2007).

Figure 2.1: Dakshin Gangotri Station in 1984[5]

Dakshin Gangotri was a permanent station until 1988-1989; after that, the station was entirely immersed in ice (Figure 2.2). This station was decommissioned in 1990 and henceforth used as a supply base. It is currently inaccessible, replaced with Maitri station (~90 km from the East coast) made operational in the same year.

Figure 2.2: Dakshin Gangotri Station after Snow Accumulation[6]

2.5.1 Maitri Station

The second research station established under the Indian Antarctic Programme (1990) is the Maitri station. In 1988, Schirmacher Hills (70° 45′52″ S Latitude and 11° 44′03″ E Longitude) was selected for the station's foundation due to its icefree rocky area. Steel stilts used to install the station into the rocky area and situated at the Schirmacher Hills base are viewed as the "Sherpa" of the mountain ranges of central Dronning Maud Land (Figure 2.3). The station has an elevation of 117 m AMSL. Maitri can accommodate a maximum of 25 persons and another 40 members in the containerized living modules placed outside of the station. The station's layout comprises a fuel storage area (Jet A1 fuel), summer camp, main building, and pump house. The amenities available at the main building include hot and cold water, a regulated power supply, cold storage, living, dining, lounge areas, and a well-equipped laboratory area. Communication with Maitri is maintained through satellite with India for voice, video, and data exchange.

Figure 2.3: Maitri Station, Antarctica[7]

2.5.1 Bharati Station

Bharati station is India's latest addition to the repertoire of research stations, located in Larsemann Hills, along the coast of East Antarctica (Figure 2.4). It is built close to the Thala fjord and Quilty Bay on Grovnes ('69° 24.41′ S, 76° 1.72′ E) 2500 km from Maitri station. This station became fully functional in March 2012 and can accommodate 15 persons through wintering and 35 during summer. It is close to the edge of the sea-peninsula junction (300 m). It has spearheaded research in understanding the geological past of the Indian subcontinent and transmitting scientific satellite data to mainland in India.

Figure 2.4: Bharati Station, Antarctica[8]

2.6 Scientific Experiments and Achievements in Indian Antarctic Stations

2.6.1 Schirmacher Hills

- Antarctic freshwater lakes' ecosystems possess low diversity of planktonic and benthic groups (Parker and Simmons, 1985). Despite low levels of resources, zooplankton biomass in lake ecosystems is the highest (Rautio and Vincent, 2006). Phytoplankton and green sulfur bacteria are some primary producers of lake ecosystems (Ng *et al.*, 2010; Keskitalo *et al.*, 2013).
- In the year 2003, a novel strain *Planococcus maitriensis* for strain S1 was isolated from Schirmacher Oasis, Maitri (Alam *et al.*, 2003). Planococcus was added to Bergey's Manual in 2015 by an Indian Scientist, described as clumps of 2 or 3 motile cocci, aerobic and catalase-positive (Shivaji, 2015).
- An underwater profile of Antarctic waters using radiometry determined that Antarctic waters indicated a higher productivity level due to high quantum penetration of UV-A and B rays (Lokabharathi *et al.*, 2001).
- The Antarctic atmosphere was studied at Maitri station to reveal dominant communities of gram-positive rods and Penicillium species (Singh and Agarwal, 2006).
- Bryophytes were isolated from Schirmacher hills revealing five genera and four families (Singh and Semwal, 2006).
- Microfauna of the Priyadarshini Lake of Schirmacher Hills was studied with the objectives; of evaluation of environmental characteristics, faunal density, and distribution (Tiwari *et al.*, 2020).
- In Antarctica, wind profiler radar operating at 54.5 MHz called Movable Atmospheric Radar for Antarctica (MARA) was installed during 2013-14 near Maitri station in Schirmacher Hills for monitoring the Antarctic boundary layer to mesosphere (Belova *et al.*, 2020).

- Climatic studies in Antarctica give a global perspective on oceanic and circulatory processes. Cooling trends assessed at Maitri station from 1990-2015 portray climate change effects (Soni *et al.*, 2017).
- Geophysical studies were conducted at Maitri station using an Absolute Gravity Meter. Reference points are set by digital relative gravimeter to monitor long term gravity daily and seasonal variations (Tiwari *et al.*, 2006c). The Magnetotelluric observations were carried out at ten stations, including eight stations in Schirmacher hills and two stations on the ice-covered region towards the south of the Oasis.
- An ice core studies carried out on an ice shelf near Schirmacher hills in East Antarctica during the summer of 2006-07. The team retrieved two shallow ice cores from different locations. During the first phase, a 75-m long core was retrieved from the ice shelf region north of the Schirmacher hills. The second phase of operations took place at another site in the ice shelf and a core of 65 m in length was obtained.
- In pursuit of abiding by the Protocol's principles on environmental protection to the Antarctica Treaty, regular environmental monitoring is mandatory for all activities (scientific and logistic). The Environmental Impact Assessment (EIA) is conducted through long-term environmental monitoring at Maitri and Bharati stations (Tiwari, 2017).
- Researchers identified the moss (*Stigonema minutum*) propagating in the vicinity of lakes near the Indian research station Maitri supporting cyanobacterial forms (Pandey *et al.*, 2004). Other samples of the Antarctic moss and soil were used to isolate the L-Asparaginase enzyme free from urease and glutaminase, used in Acute Lymphoblastic treatment Leukemia (ALL) (Ashok *et al.*, 2019).
- Ayurvedic food supplements on immunoglobulin production, lipid profile, lean body mass index, mid-arm circumference, blood pressure, pulse rate proved effective in response to

stresses generated by the Antarctic environment (Bansal *et al.*, 2013).

2.6.2 Larsemann Hills

- The digital elevation models (D.E.M.) were designed for Larsemann Hills and Schirmacher Hills, East Antarctica, using a multi-temporal RAMP, Cartosat-1 (Indian satellite) ICESat, and ground reference data (GPS) (Jawak and Luis, 2013).
- Geospatial mapping of vegetation using very high-resolution WorldView-2 data and high-resolution satellite imagery for object-oriented mapping of supra-glacial debris in the Antarctic environment. Shorelines of more than 100 lakes in Larsemann Hills and ten lakes in Schirmacher Oasis were mapped as reference data to validate algorithms for semi-automatic extraction of the lake features using satellite data (Jawak *et al.*, 2016b; Jawak *et al.*, 2016c; Jawak *et al.*, 2016a).
- The contribution of terrestrial biology studies by Indian scientists has emerged from Larsemann Hills and indicated by (Singh and Nayaka, 2017) that assessment of terrestrial biodiversity is a prospect of biotechnology in Antarctica.
- Geologists and glaciologists chose Prydz Bay Region to evaluate the tectonic evolution of the ortho and para-gneisses exposed on different islands. This study coupled with glaciological studies of Polararboken and Dalk glaciers in a long-term monitoring program to assess the response of EAIS (East Antarctic Ice Sheet) to climate change (Dasgupta and Sengupta, 2003; Pandit and De Wall, 2014).
- A state-of-the-art advanced ground station is established in Antarctica (Bharati station) to receive data from IRS. Satellites and transfer to NRSC in India (Apama *et al.*, 2014).
- Space Application Centre of Indian Space Research Organization providing near-real-time sea ice advisory to NCPOR since 2012, using data from Indian satellites, namely, RISAT, Resources at-2, AWiFS, LISS-III, LISS-IV, Saral-Altika and foreign satellite data and products for safer ship navigation to Antarctica (Gupta *et al.*, 2019).

- Climate effects on the carbon dioxide concentrations in the atmosphere above Bharati station carried out by National Remote Sensing Centre (NRSC) installing sensors which gave an insight into the effects of greenhouse gases (Pathakoti *et al.*, 2018).

2.7 Antarctica Specially Managed/Protected area (ASMA & ASPA)

The entire landmass of Antarctica is not open for access. Certain areas are cornered off, restricted or managed by party nations under the guise of the Madrid Protocol of 1991, termed Antarctic Specially Protected Area (ASPA) and Antarctic Specially Managed Area (Harris, 2002; Bauer, 2013).

Three sites out of which, India exclusively manages ASPA-163 (Dakshin Gangotri) and partly supervises ASPA-174 (Stornes, Larsemann Hills) and ASMA-6 (Larsemann Hills).

3

LIFE IN ANTARCTICA

3.1 Health and Acclimatization

Exciting research horizons have been surpassed over the past 65 years on the continent. Cross-disciplinary approaches could give the highest pay-offs in terms of Antarctic environmental insights (Gutt *et al.*, 2010). One cannot merely drop by for a visit to Antarctica. Humans visiting must be in the best possible physical and mental health to endure Antarctica's harsh climatic conditions and the sea voyage. In this regard, every year, successful candidates after defending a proposal (for scientists) or clearing written tests and interviews (for logistics personnel) and after that medical assessment at India's reputed medical centre for acclimatization training deputed at Auli (17,000 feet a.m.s.l), (Uttrakhand State) Indo-Tibetan Border Police (ITBP) camp (Figure 3.1). The members of the Antarctic expedition prepare for expeditions in advance. Following precise instructions from the NCPOR and medical experts, oral health and care must be fastidious before the journey.

Fig. 3.1: Acclimatization Training at Auli[9]

3.2 Clothing

Figure 3.2: Protective Overall[10]

Once declared 'fit' both physically and mentally, expedition members must pack necessary cosmetics like UV protective creams, goggles, and personnel clothing. During the voyage, 'Extreme Cold Weather' (ECW) outfits, i.e., overall, gloves, goggles, balaclava, mittens, etc., will be issued (Figure 3.2). Without this apparel, the body cannot sustain itself, and survival becomes difficult. Expedition members who fly from Cape Town to Antarctica carry ECW outfits and flights once issued rom NCPOR.

3.3 Working Life

Figure 3.3: Protective Overall[11]

Working in the Antarctic continent should not be taken lightly. While the Northern hemisphere experiences winter, the Southern hemisphere undergoes the summer season. The climate is harsh throughout the summer (Figure 3.3). A cold sub-zero breeze with the capacity to freeze blood makes it near impossible to work continuously outdoors. Antarctica is a very unpredictable continent to survive and conduct experiments. The once-in-a-lifetime experience of staying in Antarctica is often a defining moment.

This seventh continent is an uninhabited land with no permanent residents. In contrast, hundreds of scientists live and work for extended durations (nearly one year). People that visit Antarctica are primarily scientists, logistics personnel, or tourists, and almost all the visitors are connected to science. The Antarctic environment is not suited to all, and one must adopt a positive approach to survive there.

3.4 Natural Barriers Whiteouts

Whiteouts are a frequent, unpredictable, spontaneous phenomenon that materializes at any point across the icy continent (Figure 3.4). Although the snow precipitation levels appear low, the intensity is relatively high. Accelerated wind speeds pick up the deposited snow and deposit it over far distances. Blizzards are another disorienting whiteout condition wherein vision can perceive no horizon and depth, devastating to flights and snow vehicles.

Figure 3.4: Whiteout Condition during Voyage[12]

3.5 Food

Foods containing high calorific value and fortified with vitamins and minerals are essential to be consumed. Various countries carry fresh vegetables via cargo ship and flights with a shelf life of 2 months (Figure 3.5). Some food items are stored at -20° C for long-term use. The lack of fresh fruits and meats may seem daunting, but the professional chefs employed at the research stations provide tasty food.

Figure 3.5: IASE through Ship[13]

3.6 Population

According to an estimate, the total population (all the existing stations) in Antarctica during the summer remains around 5000 while wintering reduces the numbers to 1000. At Maitri station, during summer maximum of 55 people stay while another 50 are floating population. During the winter season, these numbers are around 25. The floating population is because of Indian expedition members who reach Antarctica by flight and stay for a few days at Maitri station to fly to Bharati station by a smaller Basleror Dornier aircraft. At Bharati station, 47 personnel can be housed on a twin-sharing basis during the summer and winter seasons and in front of the station lays a shelter/emergency camp that can accommodate an additional 25 persons.

4

ANTARCTIC ENVIRONMENT

4.1 Concept of Environment

The term "Environment" is derived from the French word 'environer' meaning to surround, envelope, encircle. This system is ruled by physical, chemical, biological, social, cultural, ethical, and economic characteristics that influence organisms. Detriments to this environment include pollution, overexploitation of land, ocean, air resources, construction and infrastructure, and human disruptions.

4.2 Dimensions of Environment

The environment can be further categorized into the following dimensions.

Physical environment (natural and constructed)**:** Soil, topography, climatic conditions, special conditions, flood plain, the potential for mudslide/landslide, subsidence, earthquake, irrigation, aerial or underground transmission lines, vegetation, wildlife, air pollution, noise pollution, water pollution.

Social Environment: Communities, public facilities, parks and recreation, commercial enterprises, service providers, the character of a community.

Aesthetic Environment: Scenic areas, vistas, natural landscapes, architectural character of buildings.

Economic status: Employed and unemployed citizens, major and minor sources of income, economic foundation in the community.

4.3 Environmental Impact

Environmental impact is an intentional or unintentional change caused to any existing characteristics of the environment or the generations of new traits to benefit or harm the environment.

Based on the severity, the Impact can be categorized as a primary or secondary impact. The primary impacts have a direct link to the proposed action. On the other hand, secondary impacts are indirectly linked.

4.4 Flora and fauna

"Which animals and plants live in Antarctica?" is the most common question of curious first-timers to Antarctica. Antarctica is devoid of vegetation except for 350 species of algae, moss, and lichen. In some regions of the Antarctic Peninsula, lush beds are observed. Lichens attached to exposed rocks of isolated mountains, including the Schirmacher Hills where "Maitri" is situated. The barren rocks host colonies of crimson, mustard, and canary-coloured lichens. Green lichens extend a few centimetres in height from the bare rock, giving a meadow effect. Over the past decade, findings of Antarctic hair grass (*Deschampsia Antarctica*) and the Antarctic per wort (*Colobanthus quitensis*) in multiple locations across the Antarctic (Komarkova *et al.*, 1985; Komarkova *et al.*, 1990; Convey *et al.*, 2012) Algal growth was also observed in orthoquartzite rocks in the dry valleys of Southern Victoria Land (Friedmann and Ocampo, 1976). Prevalent harsh conditions, like the strong, drying winds, are more detrimental to growth than low light exposure available in rock.

Figure 4.1: Seal near India Bay[14]

Antarctica is not connected to any landmass; hence there is no migration of terrestrial vertebrates. Migratory vertebrates are present and vacate the continent during the winter.

These life forms derive nutrients from the ocean. A majority of the whales feed on krill and other marine life—a diverse species habitat of seals and birds recorded there (Figure 4.1). A positive correlation was seen between Antarctic krill and Antarctic fur seal densities and macaroni penguin populations (Hunt Jr *et al.*, 1992). The second-largest mammal on the planet besides humans is the Crab-eater seal *(Lobodon carcinophagus)* found in the Antarctic. The other seals include Ross seals, Leopard seals, and Weddell seals, and their distribution depends on pack ice cover (Bester *et al.*, 2002). The population of krill is said to outweigh that of humans. Antarctica krill *(Euphasia superba)* is the keystone species of the Antarctic marine food web, and its response to climate change, as monitored using a temperature-dependent krill growth model (Pinones and Fedorov, 2016).

Figure 4.2: Emperor Penguin near India Bay[15]

The most emphasized inhabitant of Antarctica is the penguin (Figure 4.2). A bird with an excellent sense of flight or fight but ironically is flightless (Ellenberg, 2017). According to the Last Glacial Maximum (LGM), certain species of penguins are benefitting from global warming such as the Gentoo penguin benefitting over chinstrap and Adelie penguins (Clucas *et al.*, 2014). Other common sea birds include the Cape pigeon (*Doption capense*), Antarctic fulmar (*Fulmarus glacialoides*), snow petrel (*Pagodroma nivea)*, skua (*Catharacta maccormicki*) (Figure 4.3), and Wilson's storm petrel (*Oceanitus oceanicus)* (Eklund, 1964; Ridoux and Offredo, 1989).

Figure 4.3: Skua at Maitri Station[16]

4.4.1 Survival of Tiny Plants and Animals

Adaptation is the only form of survival in the severe conditions of Antarctica. Minute invertebrates such as mites, ticks, Cryptopygus springtails, tardigrades nematode worms freeze during the winter months (Wharton, 2003; McGaughran *et al.*, 2019). In order to survive, the metabolic rate is slowed coupled with the assistance of antifreeze capabilities inducing activity on thawing (Baust and Lee Jr, 1983).

The Southern Ocean harbours copious life forms from microscopic plankton to macroscopic whales (Smetacek and Nicol, 2005). A food chain determines species' existence and all associated dependent life, for example, producers, prey, and consumers.

Figure 4.4: Snow Petrels[17]

4.5 "Nunataks"- Shelter for plants and animals

"Nunatak" are outcrops of Transantarctic mountain rock peeking through the glaciers. These Nunataks offer protection to birds, such as nesting snow petrels (Figure 4.4). On the other side, Nunataks gives insight into geological studies of the dynamism of ice interactions with rock over the past millennia.

4.6 Environmental Issues

4.6.1 Environmental Awareness Past and Present

In the years preceding the 1960s, whale (Figure 4.5) and seal (Figure 4.6) populations depleted considerably due to hunting practices for meat and oil (Kock, 2007; Zerbini *et al.*, 2019). At that time, there was little or no concern about environmental conservation. After the consumption of food, the leftovers and waste were left unaccounted for. In the 1980s, Antarctic stations burnt waste in open fires or dumped the waste in the ocean (Bargagli, 2008). These antiquated practices have resulted in the perilous fate of these Antarctic species.

Figure 4.5: Whale nearby Antarctica Region[18]

4.7 Unauthorized Fishing and Present Status

Until recently, the fishing industry has not been held accountable for the overexploitation of marine resources. Certain whaling/hunting practices are deeply rooted in culture and tradition, particularly in Norway (Halverson, 2004). The Norwegians invented the exploding harpoon, and its kill streak haunts the fisheries industry to date.

Figure 4.6: Antarctic Fur Seal swimming[19]

Between 1904 and 1966, over 175,000 marine animals were killed (Pangerc, 2010). Whale meat is considered a delicacy in the East, especially in Japan (Manderson and Akatsu, 1993). Japanese whalers have been exploiting a clause in the Convention for Regulation of Whaling (ICRW) for over two decades (Sand, 2008). Antarctic krill has been harvested by the Japanese (and a small amount by Russia) in staggering quantities (~10,000 tonnes) (McElroy, 1984). During the year 2017, approximately 237,000 tonnes of krill have been reported within the designated Convention on the Conservation of Antarctic Marine Living Resources (CCAMLR) areas (https://www.ccamlr.org/en/system/files/00%20KRI48%202018.pdf, accessed on March 31, 2021). The highest krill catch was recorded in 1982 with over half a million tonnes (https://www.asoc.org/storage/documents/Meetings/CCAMLR/XXIII/ASOCKrill paper.pdf, accessed on March 31, 2021). Norway, Korea, and Japan fished the top portion of krill stock. CCAMLR regulated long-line fishing for fin-fishes in the Southern Ocean due to collateral damage of seabirds ensnared in the operation lines (Kock, 2007).

Off-shore fishing and small-scale fishing by tourists are a severe threat to the marine life of Antarctica. Antarctic fisheries in 2000-01 (1 July-30 June) reported a landing of 112,934 metric tons (Szpalski, 2006). Unregulated fishing, particularly Patagonian toothfish, is a severe problem (Collins *et al.*, 2010). The CCAMLR determines the recommended catch limits for marine species (Fabra and Gascon, 2008). Licensed fisheries targeted Patagonian toothfish *(Dissostichus eleginoides)* and Antarctic toothfish *(Dissostichus mawsoni)* in the Southern Ocean and caught 13952 and 4345 fishes, respectively, in the year 2017 (https://www.ccamlr.org/en/fisheries/toothfish-fisheries, accessed on March 31, 2021).

5

TOURISM AND ITS IMPACT ON ENVIRONMENT

5.1 Tourism Attraction

Considering Antarctica's journey is tumultuous, people are excited to visit (Figure 5.1). Tourists travel to the South pole, the most isolated continent on the planet, either through a long uncomfortable ship voyage or by an expensive flight. All visitors

Figure 5.1: A party of skiers arrives after traversing to the South Pole[20]

are subjected to unpredictable weather patterns that can turn erratic in seconds. Antarctica attracts scientists, commercial exploiters, and tourists. In the past, after the formation of the Antarctic Treaty, mining had entirely ceased making tourism the fastest growing economy associated with Antarctica. This decimates hope that the continent will remain a free and unsullied place of international cooperation.

The International Association of Antarctica Tour Operators (IAATO) stated that the number of tourists who visited Antarctica, a total of 73991 visitors, travelled to Antarctica between October 2019-April 2020 (https://iaato.org/wp-content/uploads/2020/06/IAATO-on-Antarctic-visitor-figures-2019-20-FINAL.pdf, accessed March 31, 2020).

An estimation of the 2016-17 season, 43,885 people were expected to visit Antarctica, around 5407 people more than the previous season (https://iaato.org/wp-content/uploads/2020/04/ IAATO -Fact-Sheet-2016- 17.pdf, accessed November 16, 2020) out of which 44,483 visited (https://iaato.org/wp-content/uploads/2020/04/ 2016-2017-Tourists-by- Nationality-Total.png, accessed November 16,2 020). Consequently, the initially anticipated numbers for 2017-18 were approximately 51,707 persons (https://iaato.org/wp-content/uploads/2020/03/IAATO-verview.pdf, accessed, November 16, 2020) and IAATO reported that 51,842 people visited (https://iaato.org/wp-content/uploads/2020/04/2017-18-Tourists-by-Nationality-Total.png, accessed March 31, 2021).

Ship-borne travel is favoured overflights to the Antarctic continent (Bertram, 2007 #10639). Some commercial activities by IAATO operators include the operation of small boats, shore landing, kayaking, mountain climbing, SCUBA diving, camping (including overnight stays), Unmanned Aerial Vehicles (UAV) etc. and these activities were undertaken in line with IAATO policies and guidelines (https://iaato.org/download/ip163-iaato-overview-of-antarctic-tourism-2016-17-season-and-preliminary-estimates-for-2017-18 /,accessed March 31, 2021). Studies indicate that over 100 nationalities visited Antarctica in 2017-18 (https://iaato.org/download/ip071-iaato-overview-of-antarctic-tourism-2017-18-season-

and-preliminary-estimates-for-2018-19-season/, accessed March 31, 2021s). One can also explore Emperor Penguin Colony visits, the South Pole ski expedition, and marathons to name a few (https://iaato.org/download/ip145-a-catalogue-of-iaato-operator-activities/). Operator activities are permitted or authorized through Environmental Impact Assessments (EIA), Advanced Notifications and, where relevant Waste Management permits are submitted in advance of the season.

Approximately 9,378 people travelled to the Antarctic on 2714 Traditional Seaborne Tourism Peninsula vessels operate at multiple voyage times to bring passengers to Antarctica. Similarly, 8 Air-Cruise Peninsula, 13 Sailing vessels/Yacht Tourism Peninsula, 4 strict cruise liners (Peninsula Region), 5 seaborne (Continental/ Ross Sea) and 4 Deep fields. A summary is appended in a report titled, "IAATO Overview of Antarctic Tourism: 2016-17 Season and Preliminary Estimates for 2017-18".

5.1.1 The Impact of Visitors

Antarctica is uninhabited, and any person is free to live, conduct experiments, or visit. From India, the two kinds of visitors are mostly the National Antarctic Program participants and tourists (commercial and private). In terms of numbers, the tourists greatly outnumber those from the National Program. The tourist shore landings occur for short durations but impact the environment vastly as compared to scientists and station support that visit sampling locations, residing at permanent research bases. The numbers of tourists that travel on-shore are staggering as compared to the scientists and logistics staff.

National Antarctic Programs prefer the use of the ship for the transportation of materials and personnel to the continent. Accidental crashes leading to oil spills in the Southern Ocean have occurred in the past have been reported (Bargagli *et al.*, 2007; Raymond *et al.*, 2017). Despite appropriate safeguards, the increase in ship numbers increases the probability of accidental spillages.

5.2 Tourism- Threats and regulations

The management of tourism in Antarctica is presently regulated by the Antarctic Treaty Parties and the International Association of Antarctic Tour Operators (IAATO) (Landau and Splettstoesser, 2007). The size of the vessel and visitor landing numbers are being monitored (Splettstoesser, 2000), and IAATO has been systematically regulating it (Haase *et al.*, 2009). The potential threat from tourists is from non-IAATO members aboard larger cruise ships with more significant numbers of on-shore landings (Bauer and Dowling, 2006; Landan, 2007).

The expedition trekking groups require attention to safety and elaborate rescue protocols (Dowling, 2006). Some smaller expeditions failed to adequately request aid from shipping or nearby national bases when (Lamers *et al.*, 2007). It has occurred in the past and is recorded by a few stations. Vehicular malfunction or destruction (ships, choppers etc.) leads to debris deposition at the crash site, adding to the total contamination of Antarctica (Behrendt, 2005).

5.3 Visitor recommendations for Antarctica and the Southern Ocean

Activities in the Antarctic are governed by the Antarctic Treaty (1959) and associated agreements referred to collectively as the Antarctic Treaty System (Ferrada, 2018). Antarctica is hence a land of peace and science (Cardone, 2019). Article 2 of the Antarctic Treaty designates Antarctica as a natural reserve (Chown *et al.*, 2017). The Protocol was enforced in 1998 and laid down a comprehensive approach to the protection of the Antarctic environment (Bastmeijer, 2009). Governmental and non-governmental tourism activities are bound by the Antarctic Treaty guidelines which do not compromise the environment (Bastmeijer, 2009)

This Guidance for Visitors to the Antarctic is intended to ensure awareness and compliance with the Treaty and the Protocol (Bastmeijer and Roura, 2004). Visitors are, of course, bound by national laws and regulations applicable to activities in the

Antarctic (Kriwoken, 1995).

Figure 5.2: Interactions of Antarctic wildlife[21]

5.4 Antarctic Wildlife Protection

Removal or interference with Antarctic wildlife (Figure 5.2) is harmful and prohibited except in accordance with a permit issued by a national authority.

- One should not use any vehicle that disturbs Antarctic wildlife.
- Visitors must not engage in activities with fauna (feeding and photography) as it could alter their behaviour compromising survival mode. Do not disturb animals, especially during mating or moulting.
- Visitors must not, jump on or disturb extensive moss beds or lichen-covered slopes.
- The use of guns and explosives is prohibited. An effective noise management protocol must be followed to avoid startling wildlife.
- Non-native species in the Antarctic are strictly forbidden (e.g. pets, plants or live poultry).

5.4.1 Restriction to Protected Areas

Certain Antarctic areas are afforded special protection because of their particular ecological, scientific, historic or other values (Coetzee *et al.*, 2017). Entry into certain areas is prohibited except in accordance with a permit issued by a national authority. Activities in and near designated Historic Sites and Monuments and certain other areas may be subject to special restrictions as well (Evans, 2011). Other compulsory requirements are:

- Awareness of ASPA areas and their restrictions on entry and activities.
- Compliance with the imposed restrictions.
- Removal or destruction of any artefact obtained from a historic site or monument is prohibited.

5.4.2 Value Scientific Research

Visitors may follow:

- Do not interfere with national or international scientific research, facilities or equipment.
- Obtain permission and reconfirmation 3 days prior to visitation to Antarctic science and logistic support facilities and comply strictly with the on-site rules regarding such visits.
- It is imperative that no interference, removal or disturbance to scientific equipment or marker posts, experimental study sites, field camps, or supplies.

5.4.3 Safety- Preparedness for extreme weather

While in Antarctica, one should:

- Ensure that your equipment and clothing meet the Antarctic standards. Remember that the Antarctic environment is inhospitable and potentially dangerous.
- Know your capabilities, and the dangers posed by the Antarctic environment, and act accordingly. Plan activities with safety in mind at all times.

- Keep a safe distance from all wildlife, both aquatic and terrestrial.
- Take note of, and act on, the advice and instructions from your leaders; do not stray from your group.
- Do not tread on glaciers or large snowfields without proper equipment and experience; there is a real danger of falling into hidden crevasses;
- Do not expect a rescue service; self-sufficiency is increased, and risks reduced by sound planning, quality equipment, and trained personnel.
- Do not enter emergency refuges (except in emergencies). If you use equipment or food from a refuge, inform the nearest research station or national authority once the emergency is over.
- Respect any smoking restrictions (Figure 5.3), particularly around buildings, and take great care to safeguard against the danger of fire. The low humidity in Antarctica causes fire hazards.

Figure 5.3: Smoking is prohibited[22]

5.4.4 Keep Antarctica Pristine

Antarctica is relatively pristine and has not yet been subjected to large-scale human perturbations. It is the largest wilderness area on earth. This is the responsibility of all to preserve its serenity. So, some restrictions are:

- Littering or disposal of garbage on land is prohibited as is open burning.
- Disturbance or pollution of lakes and streams is prohibited. Proper Protocol must be employed when disposing of materials at sea.
- Defacing rocks or buildings is unacceptable.
- Collection or tampering with biological, geological specimens or human-made artefacts as a souvenir, including rocks, bones, eggs, fossils, and parts or contents of buildings, is prohibited.
- Do not vandalize buildings, whether abandoned or unoccupied or emergency refuges.

5.5 Guidance for those Organising and Conducting Tourism and NonGovernmental Activities in the Antarctic

Antarctica is the largest wasteland on earth, comparatively unaffected by human activities (Clarkson, 1997), and needs to be specially protected. Furthermore, it is physically remote, inhospitable, unpredictable, and potentially dangerous (Rintoul *et al.*, 2018). Safety features and planned activities should be undertaken by those who venture to the Antarctic. The activities must be in accordance with the Antarctic Treaty (1959) and associated legal instruments, referred to collectively as the Antarctic Treaty System (ATS) (Beck, 1994). These include the Convention on the Conservation of Antarctic Seals (CCAS' 1972), the Convention on the Conservation of Antarctic Marine Living Resources (CCAMLR' 1980), along with the recommendations and implementations by the Consultative parties under the Treaty (Messer and Breth, 1991).

In 1991, the Consultative Parties adopted the Madrid Protocol under the umbrella of the Antarctic Treaty (French, 1999). The Protocol was enforced in 1998 and covered the principles, procedures, and obligations for a comprehensive approach to the protection of the Antarctic environment (Tin *et al.*, 2009). It ensures that human activities, including tourism, do not have adverse effects on the Antarctic environment (compromising its scientific and aesthetic values) (Hughes and Convey, 2014).

All activities are to be planned and an environmental impact assessment is to be conducted (Kriwoken and Rootes, 2000). Organisers must be made aware that "activities shall be modified, suspended or cancelled if they result in or threaten to result in impact upon the Antarctic environment or dependent or associated ecosystems."

The parties responsible for coordinating and executing tourism and nongovernmental activities must fully comply with national laws and regulations that implement the Antarctic Treaty system, and international law agreements on environmental protection, pollution, and safety related to Antarctic Treaty Area. It is necessary to impose upon organisers and operators the Protocol on Environmental Protection and its Annex, as they have not yet been implemented in national law. A comparative Table 5.1 depicts the difference between well-managed and ill-managed tourism.

Table 5.1: Comparison between a well-managed and ill-managed tourism enterprise.

Well managed tourism enterprise (Pros)	**Ill-managed tourism enterprise** (Cons)
Funding outreach programmes to create awareness on saving Antarctica	Reduction in scientific importance
Raising awareness of the need for environmental preservation and protection	Risks to self and others
Public interest and forums- Help promote science in Antarctica as well as open a discussion for geopolitics	Monetary Incursion

Well managed tourism enterprise (Pros)	**Ill-managed tourism enterprise** (Cons)
Peaceful conduct between visitors and resident wildlife in Antarctica	Safety and search parties
Continuous support for continual environmental monitoring programmes	Potential interference of tourists with scientific experiments

5.5.1 Future risks with the rise of capitalist tourism (current scenario)

- Increased diversity in adventure tourism and its upward trend
- The increased amount of high-risk activities
- Expansion of area for activities
- Regulation and capping the number of visitors per annum
- Private tourist facilities, including airstrips and commercial flights
- A lack of up-to-date commonly-held information about the scale and details of activities

Therefore, there is a crucial need to monitor and assess the environmental sites to mitigate the grave impacts of anthropogenic activities like tourism and to assess the future outcome if the change is not made. These studies should be conducted immediately and must be incentivised. The information generated should be available for tourism parties to create awareness.

6

ENVIRONMENTAL PROTECTION POLICIES

6.1 Antarctica Treaty

Twelve governments were the initial signatories of this treaty, namely Argentina, Australia, Belgium, Chile, the French Republic, Japan, New Zealand, Norway, the Union of South Africa, the Union of Soviet Socialist Republics, the United Kingdom of Great Britain, and Northern Ireland, and the United States of America (Franklin, 1961).

6.2 Antarctica Treaty Summary

To protect Antarctica from an international incident or discord between nations, it is freely available for research only on a peaceful basis to all of mankind (Hayton, 1960). There are many acknowledgements to science's contributions from peaceful international cooperation in Antarctica (Berkman and Young, 2005). This has helped establish a foundation of freedom for scientific exploration promoting different fields to further the progress of mankind (Franklin, 1961). This treaty is an embodiment of peace and harmony in the utilization of Antarctica's resources reflected in the Charter of the United Nations;

- Militarization of Antarctica is not permitted unless personnel and/or equipment are used for peaceful purposes.

- Freedom of scientific investigation
- Exchange of scientific plans between signatory nations and personnel whenever practical.
- No activities are to be undertaken by any nation affecting claims to sovereignty of any part of Antarctica. All territorial claims are void for the duration of the Treaty.
- Nuclear explosions and nuclear waste disposal are banned in Antarctica.
- The Treaty applies to all land and ice shelves south of 60° South, but not to the seas.
- All Antarctic stations and all ships and aircraft supplying materials and manpower to Antarctica shall be accessed by inspectors from any Treaty nation.
- Observers and exchange scientists shall follow their national jurisdiction regardless of visitation to international stations. National laws are non-applicable to stations or areas, but only to the conduct of citizens of respective countries.
- Nations will conduct regular meetings to further augment the treaty. Attendance to meetings is limited to countries engaged in substantial scientific research activity in Antarctica. Unanimous approvals are necessary for any new measures to be endorsed.
- All Nations ensure that unsolicited anti-treaty activities that are harmful to Antarctica are not allowed.
- Disputes between treaty nations, if not settled by agreement, shall be determined by the International Court of Justice (ICJ).
- The modification of the Treaty may be conducted by unanimous agreement of all parties. After 30 years, any consultative Party may convene a conference to review Treaty operations. The majority votes shall declare any modifications.

- Any nation that signs the treaty has an obligation to legally ratify it. Members of the U.N. and any countrymen may join if invited by the Treaty Nations. All notices and records are deposited with the Archives of the United States of America.
- The Treaty has been translated into four languages (English, French, Russian and Spanish).

6.2.1 Environmental Protocol

The Protocol on Environmental Protection to the Antarctic Treaty was signed on 4th October 1991 and entered into force on 14th January 1998; this agreement provides for the protection of the Antarctic environment through six specific annexes.

1. Environmental Impact Assessment (EIA)
2. Conservation of Antarctic fauna and flora
3. Waste disposal and waste management
4. Prevention of marine pollution
5. Area protection and management
6. Liability arising from environmental emergencies

It prohibits all activities relating to mineral resources except scientific research.

6.2.2 Signatories of Antarctica Treaty

The Antarctic Treaty established the legal framework for Antarctic management. The consultative status was ordained to 29 party members with 24 reserved as non-consultative (Calder-Steele *et al.*, 2013). These consultative parties are rooted in the decision-making processes of Antarctica which contains the original 7 claimants to the Antarctic continent (Hayashi, 1986). The seven claimants to Antarctica were Argentina, Australia, Chile, France, New Zealand, Norway, and the UK (Dodds and Hemmings, 2013). The Russian Federation denounced all claims to the sovereignty of Antarctica (Fogarty, 2011) along with the United States, who shared similar views (Blackie, 2016). The Antarctic Treaty

Consultative Meetings (ATCM) convenes annually and provides a platform for international negotiations with the Antarctic administration in terms of commitments, concessions and strategies (Hanevold, 1971).

On the other hand, non-claimant consultative nations are with year of obtaining consultative status - Belgium (1961), Brazil (1983), Bulgaria (1998), China (1985), Czech Republic (2014), Ecuador (1990), Finland (1989), Germany (1981), India (1983), Italy (1987), Japan (1961), South Korea (1989), Netherlands (1990), Peru (1989), Poland (1977), Russian Federation (1961), South Africa (1961), Spain (1988), Sweden (1988), Ukraine (2004), United States (1961) and Uruguay (1985); non-consultative members, with year of accession in parentheses, are - Austria (1987), Belarus (2006), Canada (1988), Colombia (1989), Cuba (1984), Denmark (1965), Estonia (2001), Greece (1987), Guatemala (1991), Hungary (1984), Iceland (2015), Kazakhstan (2015), North Korea (1987), Malaysia (2011), Monaco (2008), Mongolia (205), Pakistan (2012), Papua New Guinea (1981), Portugal (2010), Romania (1971), Slovakia (1993), Switzerland (1990), Turkey (1996), and Venezuela (1999).

6.2.3 Treaty Nation

To date, a total of 54 countries have signed the Treaty, few of them or original signatories, few are of consultative status, and few of them pertain to Acceding State.

6.3 Madrid Protocol

The Madrid Protocol is the legal backbone of the Antarctic Treaty, which reiterates the responsibility of party nations to protect the Antarctic environment (Tamm, 2018). It is in constant ratification regarding the upgrade of agreed measures and recommendations (Blay, 1992). The Protocol on Environmental Protection to the Antarctic Treaty is placed as Annexure-1.

6.3.1 Protocol summary

- Antarctica is regarded as a natural reserve to promote peace and scientific discovery while prohibiting mineral activities.

The Antarctic Treaty Protocol streamlines guidelines for all proposed activities (Kirkham *et al.*, 2020).

- The Protocol requires that all proposed activities and their associated environmental impacts be presented to the ATCM through the Committee for Environmental Protection (Sánchez and McIvor, 2007).
- Addresses contingency plans to prevent environmental disasters on the Antarctic content and the surrounding ocean (Poland *et al.*, 2003).
- Supports a liability annex to safeguard Antarctica from potential threats of environmental damage (Blay and Green, 1995).

6.4 Annex to the Protocol

The Protocol is also accompanied by Annexe that detail specific measures and procedures related to:

6.4.1 ANNEX I: Environmental Impact Assessment (EIA)

The proposed activities are categorized according to their degree of impact on the Antarctic environment (Montarroyos *et al.*, 2019). Based on the degree of impact, either an Initial Environmental Evaluation (Najafi *et al.*) or a Comprehensive Environmental Evaluation (CEE) is prepared. As per procedure, the document is reviewed by the Committee for Environmental Protection and the remaining consultative parties (https://ats.aq/documents/recatt/Att605_e.pdf, accessed November 19, 2019).

6.4.2 ANNEX II: Conservation of Antarctic fauna and flora

The proposed activities must not interfere/harm/disturb Antarctic species thereby affecting their productivity (Figure 6.1). A permit is required to perform the proposed activity if there is a likelihood of the introduction of invasive species (Hughes and Convey, 2014). It also discusses the gambit of invasive species (Smith, 1996).

Figure 6.1: Mandatory soil sterilization[23]

6.4.3 ANNEX III: Waste disposal and management.

This annex defines the routes of disposal, treatment, and removal of wastes generated in Antarctica (Smith and Riddle, 2009). Human waste, for example, must be incinerated, and the ashes or sludge collected in spill-free drums and carried back to the respective nations (Figure 6.2). Need for improvements to current waste technology are required to deal with the presence of harmful contaminants such as PCBs, polystyrene packaging beads, and pesticides that are currently prohibited in Antarctica.

Figure 6.2: Incinerator used for Excreta Burning[24]

6.4.4 ANNEX IV: Prevention of marine pollution

Regulates ship discharge and disposal of ship wastes into the ocean. It refers to the International MARPOL 73/78 convention (Harris and Meadows, 1992). MARPOL prohibits plastic disposal into the sea (Derraik, 2002).

6.4.5 ANNEX V: Specially protected areas and their management

Categorization of areas into specially protected and managed areas (Figure 6.3) wherein the entry is restricted to permit holders (Goldsworthy and Hemmings, 2009). Historic sites and monuments are also protected under this annexe (Barr, 2018). Any damage, removal or interference is prohibited in these sites (Tiwari, 2006).

Figure 6.3: Historic Site and Monument (HSM- 44)-
Dakshin Gangotri –Protected Area[25]

6.4.6 Annex VI: Liabilities for Environmental Damage

Damages often arise during routine logistics activities, scientific program research, tourism, and other governmental and commercial activities in Antarctica. Hence, proposed activities must be discussed in advance to prevent environmental damage (Johnson, 2006). The proposals for plans, prevention of potential disasters and emergency contingencies must be considered.

6.5 Environmental Regulations in Nutshell

With regard to the Antarctic species and their protection, the following regulations have been enforced: -

- Removal, capture, collection, extermination of any member from Aves or Mammalia is prohibited without permit possession (Areas *et al.*, 2001). The permit will only be granted on a strong case of scientific research (Blay, 1992).

- The introduction of soils on fresh foods may contain non-native plant seeds, fungal spores and invertebrate eggs. This is one effort to control invasive species propagation (Richards, 2015).
- ASMA's and ASPA's are areas of scientific interest and must be protected for their inherent values (Neumann, 2020).
- Due to gross seal extinction, protection was enforced through the Agreed Measures for the Conservation of Antarctic Fauna and Flora (Antonello, 2014) and the Convention for the Conservation of Antarctic Seals (CCAS) (Constable *et al.*, 2000) although recently, an appeal for the de-listing of Antarctic fur seals was proposed to ATCM in June 2006 as the species were not in any immediate danger (Jabour, 2008).
- The Convention on the Conservation of Antarctic Marine Living Resources (CCAMLR) has incorporated precaution management to limit commercial overexploitation of marine living resources in the Southern Ocean (Fabra and Gascon, 2008).
- Sea disposal of wastes and ship discharges are strictly regulated (VanderZwaag, 2010). Mining and any related activity for commercial gain are banned (Floren, 2001).

6.5.1 Antarctic travel permit

Visitation to the Antarctic requires the procurement of a travel permit issued by the Antarctic Treaty State government. The permit is granted based on following the stipulated rules with a cap on the number of permits issued per year.

6.5.2 Ship waste

Ocean discharge of ship wastes (Figure 6.4) is illegal in Antarctica and must be held until port reception facilities outside the 60° S Treaty area (Bou Franch, 1996). The retention of oily wastes on-board is necessary when in Southern Ocean territory (Blay and

Green, 1995). An oil book record is used to monitor the number of oil discharges mandated by the International Maritime Organization (IMO) (Butt, 2007). It can be categorized into the following: sewage, food waste, and other waste.

Another waste generated on ships is stored onboard until it returns to a port outside Antarctica (Hampson, 2001). The use of shredders for processing glass and metal, compactors to bale plastic, and marine incinerators to burn paper (Roseblade, 1996).

Figure 6.4: Waste scheduled for ship transportation[26]

6.5.3 Sewage

In accordance with the Madrid Protocol, domestic ship sewage is biologically pre-treated prior to discharge into the open ocean (Smith and Riddle, 2009). In light of system failures, discharge of untreated raw sewage and food waste is not permitted within 12 nautical miles from the shore (Barnes and Conlan, 2007).

On the other hand, Article 5 states that food waste must be macerated and passed through a mesh (<25 mm) (Harris and Meadows, 1992). If the ships are within 12 nautical miles offshore, the oily mixtures of food waste are placed into holding tanks until discharge (Hampson, 2001). Larger food wastes and wrappings incapable of shredding are subjected to freezing and storage followed by disposal at port reception facilities outside the confines of Antarctica.

6.5.4 Waste from Antarctic bases

Disposal of wastes (physical, chemical, biological/food) onto Antarctic land/sea/ice-shelf/sea-ice and open burning activity are all banned. Instead, wastes (Figure 6.5) need to be segregated at the source and processed using compaction and shredding equipment for volume reduction followed by removal from the continent. The garbage generated at Antarctic bases is packaged and shipped off Antarctica and disposed of by licensed waste contractors. As permitted under the Environmental Protocol, the discharge of sewage and food waste into the sea can occur after biological treatment and physical breakdown.

Figure 6.5: Dispersed station debris[27]

On the other hand, select bases discharge waste into ice pits. These wastes are not removed from Antarctica because of the sheer health risks involved in shipping large quantities over long distances. The extent of the environmental effects due to the release of sewage and food effluent into the nearshore marine environment is an ongoing environmental monitoring undertaking (Tin *et al.*, 2009). Such effluents have a minor and local impact on the Antarctic environment (Stark *et al.*, 2006). To reduce these impacts, an increasing number of bases have installed biological sewage treatment plants equipped with UV accessories (Gröndahl *et al.*, 2009).

Administrating annual audits on waste generated per annum at research stations, field camps and ships by National Programs can hope to shed light on improvements to manage waste effectively.

6.5.5 Dogs

Greenland huskies were once pro-choice for Antarctic land transportation before the dawn of mechanized equipment (Pratt, 1960). Many stalwart expedition crew members regarded sledge dogs as close companions enriching their Antarctic experiences (O'Higgins, 2018 #10755).

In 1991, claims that the canine distemper virus spread from sledge dogs to sea lions caused concern with the Antarctic Treaty System (Kerry and Riddle, who decreed that "Dogs shall not be introduced onto land or ice shelves and dogs currently in those areas shall be removed by April 1st, 1994". On the other hand, the removal of huskies from Antarctica was thought to be a political ploy (Huston, 2001).

6.5.6 Mineral Resources

Since 1988, two international conventions have protected the Antarctic from commercial mineral exploitation. The first being, the Convention on the Regulation of Antarctic Mineral Resources Activities (CRAMRA) was adopted at the final ATCM in June 1988 (Beck, 1989). Australia and France proceeded to repudiate their standing regarding this mineral convention and reiterated that the Protocol on Environmental Protection is improved upon (Blay, 1992). It was then signed in Madrid in 1991 again by all 26 voting parties of the Antarctic Treaty protecting the continent for the next 50 years against all mineral activities (Treaty, 1959). The convention requires that all member countries enact national legislation to make it international law. The mineral regime is evolving to endorse the opinions of ATS participants (Joyner, 1988).

7

ENVIRONMENT MANAGEMENT AND PROTECTION APPROACH

7.1 Waste Management in Past

The use or consumption of most matter produces waste. The total Antarctic population per annum is comparatively less but is still a threat to Antarctica's untarnished nature. Over the past century, little or no weightage was offered to maintain the environment. During 1980-1990, waste was disposed into the sea or a base adjacent to the designated landfill or burnt (Bargagli, 2008). The abandonment of larger obsolete vehicles and equipment occurred at different locations (Stark *et al.*, 2006). For instance, bulk debris deposited on sea ice would eventually be deposited into the sea through ice melting (Kanfoush *et al.*, 2000).

Inland waste is usually discarded in a crevasse or left as such. Instances of oil drums being blown by wind resulting in oil spillages have been documented (Curl *et al.*, 1992).

The clean-up of past and present contaminated work sites is a prerequisite of the Madrid Protocol (Snape *et al.*, 2001). On the condition that the procedure has no impact on the natural environment or causes destruction/removal/displacement of historic monuments and sites (Hodgson-Johnston *et al.*, 2017).

7.1.1 Present Approach

Environmental audits are crucial towards managing environmental norms at bases (on land and at sea), to determine

Figure 7.1: Air Emission from Ship[28]

whether base operations and activities impact local environments (Bastmeijer and Roura, 2008). Some anthropogenic activities cause damage to its immediate surroundings for example vehicular exhaust, wastewater and other types of waste from Antarctic bases (Tin *et al.*, 2009).

Fossil fuel is used to power vehicles, ships generators and incinerators (Figure 7.1). The utilization of fossil fuels has the potential for air pollution (Chow and Watson, 2002) and oil spills (Aislabie *et al.*, 2004). To overcome and manage this situation an alternate source of energy is required as a source of energy and to do so some bases like, Finland, Belgium, Spain, the United Kingdom, Australia, Germany, France, India, South Africa, the United States, Russia and Japan are experimenting and generating electrical power from a wind turbine (Tin *et al.*, 2010). Flarvesting a natural energy source such as wind is an obvious substitution. Increased awareness has made visitors attune to protecting the Antarctic environment.

7.2 Potential of Commercial Mineral Pollution from Deposits

The current extraction of vast mineral deposits in Antarctica is unprofitable (Cullen, 1994). In addition, fast-moving glaciers could result in catastrophic glacial earthquakes (Nettles and Ekström, 2010) creating further pollution.

7.2.1 Coal

In the International Geophysical Year (IGY) of 1957-1958, coal deposits were discovered along with the Transantarctic mountain ranges with high rank and non-caking characteristics with high moisture and oxygen contents (Schopf and Long, 1966). The

National Science Foundation (NSF) of the United States of America commissioned analyses of Mount Gran coals and the results depicted low volatility, 9-15 % ash content and a high calorific value (Mulligan, 1963).

Better reserves are available elsewhere in the World. The Prince Charles Mountains offered comparatively better coal reserves, but the feasibility and economic viability of acquisition proved futile (Zhang and Talalay, 2021).

7.2.2 Iron ore

The largest iron deposit was found in the Prince Charles Mountains, East Antarctica containing Precambrian banded iron formation (jaspilite) whereas Mount Rucker deposits are of a lower Fe grade (Congress, 1989).

Surface rocks contain high amounts of iron ore trapped under ice containing only 35% iron. The Antarctic runoff obtained from glacial weathering produces filterable iron 6-81 kg km-2 a-1 contributing to the Southern Ocean results in the boost of productivity (Hodson *et al.*, 2017). Iron meteorites were also found in Antarctica as a result of asteroid impacts (Wasson, 1990)

7.2.3 Chromium

Theoretically, the Durek Massif East Antarctica is a possible source of the element Chromium.

7.2.4 Oil and gas

According to prior research, rock formations in Antarctica contain oil and gas reserves (Ivanov, 1985). Fortunately, these reserves remain untouched as mining in the Antarctic continent and associated sea bed are prohibited (McColloch, 1992). Undoubtedly, it would be an expensive resource in the market. Considering the vast reserves of oil in the world today, the need to exploit Antarctica for oil is feeble.

7.2.5 Mining

Since time, in memoriam mining has not been conducted in Antarctica.

7.3 Environmental Protocol Restrictions

During the Antarctic Treaty signing in 1959, the exploitation of resources was not discussed (French, 1999). Key issues were highlighted in the late 1980s leading to the Madrid Protocol on Environmental Protection to the Antarctic Treaty signed in 1991 institutionalizing a ban on mining till 2041 (Vidas, 2002).

7.4 Environmental Management at Indian Permanent Research Stations - "Maitri" and "Bharati"

Maitri is India's 'year-around' functional research base, positioned on Nunatak Vassfjelleton, Schirmacher Hills, along Princess Astrid Coast, East Antarctica (Tiwari and Nayak, 2007). Maitri was built into a barren rock covered with ice during the winter months and exposed during the summer months to reveal small lakes (Tiwari and Kulkarni, 2006).

According to the Madrid Protocol on Environmental Protection to the Antarctica Treaty a continuous environmental monitoring and impact assessment of base operations and activities is mandatory and must include:

7.4.1 Air Environment

- Monitoring of natural radioactivity level around Maitri
- Assessment of indoor radon levels in various huts of Maitri
- Assessment of natural background radiation levels, heavy metals in air.
- Air quality monitoring baseline and from base operations and activities, to determine SPM, RSPM, SO_2, NOx, Methane, Hydrocarbon, CO_2 and CO (Figure 7.2).
- Micrometeorological data collection along with radiosonde data for calculation of mixing height and ventilation coefficient.
- Validation of Air Quality Prediction Model.

Figure 7.2: Air Monitoring at "Maitri" Station Using High Volume Sampler[29]

7.4.2 Water Environment

- Analysis of physicochemical parameters of surface water bodies (Figure 7.3)
- Analysis of physicochemical parameters of the wastewater treatment system
- Performance evaluation of wastewater treatment system (Rotating Biological Contactor)
- Efficacy improvisation of the wastewater treatment system
- Application of tertiary treatment to present wastewater treatment plant
- Characterization of sediments sample from lake and wastewater collection pond

Figure 7.3: EIA Laboratory at Maitri Station[30]

7.4.3 Land Environment

- Analysis and characterization of soil samples collected around the Maitri station for various physicochemical properties and evaluation of contamination.
- Assessment of oil spill and characterization of affected oil spilt soil samples.

7.4.4 Solid Waste

- Assessment and quantification of hazardous and non-hazardous waste and its management.

7.4.5 Noise Environment

- Frequency distribution of sound pressure levels analysis of various types of machinery.
- Day-night noise levels measurement in and around Maitri.

7.4.6 Biological Environment

- Study of distribution and abundance of microfauna with associated algal and moss communities
- Analysis of drinking water sources and other lakes for phytoplankton and zooplankton
- Study of microfauna of Priyadarshini Lake
- Study of nutrients and phytoplankton in coastal and offshore waters
- Effects of ultraviolet radiation in the terra of the Schirmacher Region of East Antarctica

7.4.7 Environmental Impact Assessment

- Baseline assessment of air, noise, water, land and biological components
- Identification and quantification of impacts of the activities during the study and proposed activities- minor, moderate, major and significant.
- Evaluation of existing environmental management measures
- Preparation of Environmental Management Plan (EMP)
- Preparation of effective environmental monitoring program.

7.5 Environmental Awareness in the Antarctic Expedition

7.5.1 Environmental impact and monitoring- Human Activities and Natural Process

Anthropogenic effects tend to affect the wholesomeness of the Antarctic environment either directly, indirectly or cumulatively (Gilpin and West, 1995). Inherent existing biological and abiotic characters of the environment must be assessed for risk of vulnerability to human interactions (De Lange *et al.*, 2010). Understanding the contributing factors to the environmental baseline is important before the initiation of any planned activity (Bargagli, 2006). After the planned activity is completed, recovery of the environment by remediation or removal needs to be initiated (Parker and Wiens, 2005).

The investigation of population dynamics influenced by seasonal and long-term changes is a requirement (Convey and Peck, 2019). Whether human influence has a substantial impact on the Antarctic environment with its natural order must be explored (Goudie, 2018). The extent of protection Antarctica can extend against human pollution is limited (Hughes and Convey, 2010).

7.5.2 Non-Antarctic Threats

Due to global industrial expansion, growing populations and demand, production and manufacturing have increased (Chenery, 1960). These processes generate contaminants that spread using the world's circulatory pathways having Antarctica as their sink (Van Den Broeke, 2004); (Rintoul, 2010). Moreover, these global contaminants are deposited via long-range transport into the Antarctic environments (Jackson, 1997); (Macdonald *et al.*, 2003); (Li *et al.*, 2016). The Antarctic continent remains physically disconnected from other continents but connected via human movements and influence (McLoughlin, 2001); (O'Reilly and Salazar, 2017).

7.5.3 Scientific Uncertainty

Gaps in knowledge about human impacts on an existing environment could significantly threaten the existence of fragile ecosystems. Hence, scientific counsel from experts in Antarctic environments is advised.

7.5.4 Need for Waste Management and Environmental Awareness

All base operations and activities along with relevant details must be logged in to an information portal so that records can be maintained, locations can be identified efficiently, and environmental disasters can be predicted and managed accordingly (Sinha *et al.*, 2019). Geospatial support enhances the management experience by pinpointing disaster areas and tracking the fastest route for first responders as well as determining the degree of impact (Council and Committee, 2007). On-field supplies and equipment must be removed unless in possession of a permit. For example, permit 2 is for the collection/disturbance of geological/ biological/man-made samples collected from an ASPA.

Wastes (human wastes, greywater, food waste etc.) generated must be stored and transported to the ship for disposal at port reception facilities or back to the respective country (Bharti *et al.*, 2016). Training on the environmental consciousness of expedition members for the prevention of the dispersal of foreign materials into the Antarctic environment is necessary (Harris and Meadows, 1992). Before embarking on-field activities, the removal of additional packaging must be undertaken to prevent unnecessary litter and to save time.

7.6 Mechanism for Controlling Alien and Invasive Species

To control species invasion, the best preventative measure is to create a source barrier (Hulme, 2006). Supplementary to this, the use of geographic information systems (GIS) and artificial intelligence (AI) could promote an effective ground monitoring system (Martinez *et al.*, 2020). Improper cargo storage and handling can be considered a point of the influx of invasive species (Shannon, 2008). Control measures by shifting the position of cargo storage to a more sterile area if known invasive species are located nearby (Lowe *et al.*, 2000). Food material is an alternative carrier and must be pre-treated prior to the voyage

(Venkatramanan *et al.*, 2017). The introduction of storage enclosures with oxidizing generators (UV or ozone) to reduce the survival of invasive species can be undertaken by research stations (Buck, 2010). These potential approaches could limit the alien introductions at various Antarctic locations.

7.6.1 System Development to Monitor Invasion

Development of an invasive species reporting platform is possible, but how can one accurately identify the species, carrier(s) and sources? Some sources may include voluminous stores, cargo, wharves and personnel clothing. The key is to have multiple personnel involved in the process of observation, recording and reporting their findings virtually.

A record (6 months) of the travel history of each visitor to the Antarctic must be made available. Often critters find their way into luggage and cargo. Entrapped species or soil fragments are found in equipment carriers, kits, packs, pockets and shoes.

7.7 Transportation and Fuel

One must take necessary precautions whilst traversing between field locations to not disturb the natural local environment. Mechanized transportation should be essentially restricted to minimize atmospheric emissions, track formations, physical disturbances, wildlife disturbance and the potential for fuel spills. It is recommended that vehicular use be restricted within permanent ice/snow and designated ice-free routes (Tiwari *et al.*, 2006a). Fuel spills are difficult to clean up and take many years to dissipate (Ansell *et al.*, 2001). Refuelling or changing oil amidst windy conditions could cause accidental spillage and hence filling prior to departure, is suggested. One should always use fuel cans with nozzles/funnels and an absorbent spill pad/drip tray. When travelling on foot, one should use established tracks and designated crossing points wherever possible. Where such tracks do not exist, the most direct route can be taken to avoid vegetated areas and delicate geological formations, such as sediments, streambeds and lake margins. These features have taken many thousands of years to form and may also be of major scientific importance.

One should extremely be cautious not to feed the wildlife. When around wildlife is quiet, move slowly, stay low to the ground and adhere to the minimum approach distances, unless authorized by a permit.

7.8 Awareness among Antarctic Visitors

To protect the environmental values of Antarctica, awareness is essential for pollution management and control measures. This establishes reflection into the management plan and human influence on the Antarctic Treaty area. The Environmental Impact Assessment studies are followed the environmental code of conduct preparation (Deegan and Gordon, 1996). Environmental awareness among the expedition members and tourists will help protect Antarctica's untouched environment (Blay, 1992). The Antarctic area is managed by the protocol on environmental protection to the Antarctic Treaty and is under peaceful governance, but is not to be taken lightly (Lord, 2020). Therefore, accountability for environmental footprints must be understood as we represent our nation in Antarctica.

8

ENVIRONMENTAL STUDIES IN ANTARCTICA: INTERNATIONAL AND INDIAN PERSPECTIVE

8.1 Environmental Diversity

Antarctica supports a diversity of life from the tip of Vincent Massif radially towards the Southern Ocean. This continent's climatic conditions are unique due to the Polar vortex that maintains frigid air parcels over the East Antarctic whereas, an excluded and much warmer West Antarctica lies beyond the vortex. This dynamic existence of climate promotes the growth of psychrophilic and psychrotolerant species. These species are endemic to Antarctica and must be protected following the Antarctic Treaty System through rectification of negative impacts. These environmental concerns comprise air, water, and soil pollutants along with biological entities. On international platforms, key concerns have been addressed and abated through environmental management policies their enforcement, whereas India has made great strides thus far.

For years, scientists explored to understand the extreme and pristine environments of Antarctica. The responsibilities are shared by sister signatories under the Antarctic Treaty system to conduct scientific research. The Madrid protocol in 1991 was imposed for the environmental protection of this continent. Therefore, Environmental Impact Assessment (EIA) has become mandatory for all party nations engaging in scientific activities.

Scientists have been on the lookout for new affordable remedies to existing issues. Climate change effects are seen in Antarctica due to aerosols' deposition, black carbon, persistent organic pollutants, greenhouse gas emissions and trans-oceanic transport, ozone hole depletion, heavy metal leaching introduction of alien species, and reduction in existing endemic populations. Juggling climate change and the search for unique solutions has been challenging but made possible through Antarctic research.

8.2 Marine and Freshwater Environments (International Perspectives)

To find solutions, one must assess the issue and availability of resources. Determination of baselines of local environments without human pressures is a necessity, and after that, annual estimations provide data on human interventions or natural phenomena on the continent.

8.2.1 Microplastics

The Amundsen Sea has reported macroplastics without a direct tie to man (Barnes *et al.*, 2010). Although plastics have reached Antarctica, the subtropical front acts as a barrier preventing most global transport to the continent (Suaria *et al.*, 2020). The threat of drifting plastic debris could potentially piggyback invasive species to Antarctica (Derraik, 2002). Macroplastics through weathering processes can be converted into ingestible microplastics detrimental to marine life. Known wastewater sources unaided do not contribute microplastics to the Antarctic Ocean (Waller *et al.*, 2017). A study wherein fibrous microplastics in marine sediments near a sewage treatment plant outfall at Rothera station has shed light on introducing contaminants via anthropogenic influence (Reed *et al.*, 2018). Furthermore, polyurethane and polyamide microplastics are locally produced by Antarctic stations (Lacerda *et al.*, 2019). Screening of Haswell seabird populations in Mirny, Antarctica, elucidated the dangers of macro and microplastics (Golubev, 2020).

8.2.2 Connection with Persistent Organic Pollutants

Microplastics have been linked to the transport of persistent organic pollutants (Caruso, 2019). The endocrine disruption of males of *Notothenia Rossii* and *Notothenia coriiceps* (marine benthic fish) was due to a combination of P.O.P.s (Ferreira *et al.*, 2019). On the other hand, bacterial communities also suffer from the influx of hydrocarbons, polychlorinated biphenyls, and heavy metals that affect the marine ecosystem's functionalities (Giudice *et al.*, 2019). Generation of P.O.P.s (PAH and P.C.B.s) from the combustion of fuel and incineration of wastes has been well documented. Besides this, trace metals' production through corrosion or ageing of ships is how these contaminants entered marine environments. Perpetuity and persistence of these pollutants have been blatantly caused by humans residing in research stations.

Another such concern includes heavy or traces of metals. Cadmium and copper have been bio-accumulated in high amounts in Antarctic marine copepods compared to studies of their Arctic counterparts (Kahle and Zauke, 2003). It is well known that ecosystem health can be assessed with the help of biomarkers. Lipid biomarkers have paved the way to understanding the endemic meltwater processes of microbial mats (Jungblut *et al.*, 2009). Harmful contaminants are often assimilated by resident organisms (natural biomarkers) close to the stations and measured by quantitative ethoxyresorufin-O-deethylase (E.R.O.D.) assays (Kennicutt *et al.*, 1995).

In 2004, esterase was monitored in *Adamussium colbecki* (bivalve Antarctic mollusk) for the effect of pesticide congeners and heavy metal concentrations (Bonacci *et al.*, 2004). Similarly, soft tissues of Antarctic clam *(Laternula elliptica)* were used as a prospective bio-monitoring tool for coastline health (Vodopivez *et al.*, 2015). Indicators for faecal pollution in Terra Nova Bay were assessed by densities of enterococci, streptococci, and other coliforms in *Laternula elliptica* when stations increased the number of expedition participants (Bruni *et al.*, 1997). Reduction of faecal coliform loads in marine environments was spearheaded in the

North Cove near Rothera Research station by installing a sewage treatment plant (Hughes, 2004).

The total P.O.P. concentrations up to 143 pg L-1 were observed firstly in Lake Carezza, followed by Lake Gondwana (Vecchiato *et al.*, 2015). The northern peninsula freshwater reported ~30 compounds capable of disturbing endocrine function (Esteban *et al.*, 2016). Sixteen PAHs were detected in sampling sites from ice-free streams on King George Island, Admiralty Bay (Szopinska *et al.*, 2019). Six lakes in Victoria Land were tested to reveal staggering types of PBDEs and P.C.B.s (Corsolini *et al.*, 2019).

P.O.P.s travel via Long-Range Transport (LRT) and distribute themselves between the abiotic and biotic phases efficiently (Negoita *et al.*, 2003). P.O.P.s ability to volatilize/re-volatilize followed by their indiscriminate deposition makes them persist in the environment (Cabrerizo *et al.*, 2013). The Antarctic content experiences precipitation commonly in the form of snow. It is predicted that climate change will increase snowfall towards the South Pole (Winkelmann *et al.*, 2012). Snow plays a pivotal role in the scavenging of P.O.P.s in Antarctica (Casal *et al.*, 2019). The increased temperature as a result of climate change has led to the release of P.O.P.s entrapped in permafrost across the continent (Potapowicz *et al.*, 2019).

8.2.3 Microplastics in Freshwater environments

Formulation of policies is the most beneficial method to curb the introduction of microplastic introduction into freshwater systems (Eerkes-Medrano *et al.*, 2015). The presence of fibrous and film microplastics in a freshwater stream on the Byers Peninsula (Antarctic Specially Protected Area) has been recently reported (Gonzalez-Pleiter *et al.*, 2020).

8.2.4 Heavy and Trace Metal

The freshwater lakes in Larsemann Hills were found to have low concentrations of noteworthy environmental trace metals with low inputs from the surrounding environments (Gasparon and Burgess, 2000). Terra Nova Bay freshwater lakes were assessed to speculate the sources of elements (Abollino *et al.*, 2004). Additionally, a baseline study was piloted to characterize freshwater lake geochemistry in Larsemann Hills by Gasparon, M. and Matschullat, J. in 2006. Freshwater lake profiles with mean concentrations and designated ranges of trace metals of Vecherny Oasis, East Antarctica, have been elucidated (Kakareka *et al.*, 2019). (Convey, 2010) have well explained the challenges faced by terrestrial biodiversity of Antarctic ecosystems.

Trace elemental sources around Ferraz station were sewage disposal, paint remnants, and fuel intermediates (Santos *et al.*, 2005). The Artigas base soils had the highest concentrates of trace metals (Zn, Pb, Cr) than sediments from freshwater and marine sources (Bueno *et al.*, 2018). The potential destabilization of soil ecosystems by high concentrates of heavy metals such as copper (Cu), cadmium (Cd), and mercury (Hg) makes these ecosystems vulnerable to pollutants (Chu *et al.*, 2019).

8.2.5 Atmospheric environments

Lichen biology allows the organism to be an excellent candidate for air pollution essays in Antarctica (Gries, 1996). A typical air pollutant is black carbon, which reduces the albedo of snow. Black carbon monitoring at McMurdo station determined a significant portion of local station emissions (Hansen *et al.*, 2001). Long and short-term variability studies on black carbon trends were conducted at Neumayer station, Antarctica (Weller *et al.*, 2013). Zangrando *et al.* (2016) detected aerosol pollution of levoglucosan and phenolics in the range of 4.1 and 6.4 pg m-3 and 7.3 and 15 pg m-3 median concentrations, respectively. These organic tracers were used to highlight areas where biomass burning occurred (Shi *et al.*, 2019).

8.2.6 Bio-remediation

The addition of a pesticide (INIPOL EAP 22) to oil-contaminated Antarctic coastal sea waters (near "Terre Adelie") showed an enhancement of saprophytic and hydrocarbon degraders (Delille *et al.*, 1998). Local station pollution management was initiated by microcosm analysis to assess the inherent abilities to reduce hydrocarbons via biodegradation fortified with nitrogen and phosphorus under the guise of bio-stimulation (Ruberto *et al.*, 2003). Alternatively, land shoreline pollution recovery can be achieved using land farming methods of controlled nutritional release in the presence or absence of water and aeration instead of natural attenuation (Snape *et al.*, 2003). Pelletier *et al.* (2004), successfully conducted in situ experiments on hydrocarbon degraders (stimulated) in inter-tidal sediment samples rich in by-products of oil spills. Remediation of the waste site was likewise carried out by exploiting Paramoera walker bioaccumulation (Palmer *et al.*, 2006). Walworth *et al.* (2007), conferred that the excess of nitrogen increased the rate of petroleum degradation in sub-Antarctic soils. Lipases could be substituted to degrade grease, fats, and lipids in domestic wastewaters Joseph *et al.* (2008), Role of Cold Active Lipases in Bioremediation). Diesel contaminated soil (Casey station) remediated using bio-pile (A bio-pile is a bioremediation technology in which excavated soils are mixed with soil amendments, formed into compost piles, and enclosed for treatment.) was accomplished by McWatters *et al.* (2016).

8.2.7 Bio-prospecting advances

Since the 1980s, National Antarctic programs have participated in the bioprospecting industry viz. agricultural, biotechnological and pharmaceutical sectors, leading to many discoveries. Organisms have a few fail-safe methods to combat the immense cold and survive Antarctica's extreme climate. One such mechanism is the use of cold-active enzymes that are highly sought after in industrial applications. Lipases and proteases are global bio-catalysts in many large-scale production processes. A lipase enzyme nucleotide sequence (triacylglycerol acyl hydrolase E.C. 3.1.1.3.) of 1469 base pairs was isolated from the Antarctic Moraxella TA 144 strain (Feller *et al.*, 1990). Antarctic soil fed into a packed column bioreactor revealed promising protease producers *(Chryseobacterium* and *Massilia)* (Wery *et al.*, 2003). Industrially significant high-weighted polymers were produced from a Gammaproteobacteria member isolated from a marine

Antarctic source at sub-temperatures (Nichols *et al.*, 2005). On the other hand, filamentous fungi from Livingston Island produced cold-active antioxidant enzymes (superoxide dismutase and catalase) (Tosi *et al.*, 2010). Candida Antarctica was found to possess the ability to convert n- alkanes into glycolipid bio-surfactants, mannosylerylthritol lipids (Kitamoto *et al.*, 2001). Another lipase gene was also cloned, expressed, and characterized from Candida Antarctica ZJB09193 and was used to produce Vitamin A esters (Liu *et al.*, 2012). *Oleispira Antarctica* genome sequencing and functional genomics have enlightened the use of cold adaptations to remedy marine oil spills (Kube *et al.*, 2013). The effects of trace metal concentrations on Antarctic bacteria's biosynthesis have been studied in the form of CdS quantum dots and nanoparticle synthesis in response to arsenic exposure (Glatstein *et al.*, 2018).

Cancer research is currently a top priority in the medical healthcare industry, and scientists look to the Antarctic. Santiago *et al.* (2012), discovered antagonistic endophytic fungi from endemic Antarctic plants, i.e., *Deschampsia Antarctica* and *Colobanthus quitensis* towards Leishmania and Trypanosoma. For the treatment of colorectal cancer, "Antarctica" is isolated from *Deschampsia Antarctica Desv.* (Malvicini *et al.*, 2018).

Four macro-algae *(Cystosphaera jacquinotii, Iridaea cordata, Himantothallus grandifolius, Pyropia endiviifolia*) have been investigated with bioactive compounds with antimicrobial and cytotoxic effects against cancer and non-cancer cell lines (Martins, R.M. *et al.*, 2018). Marine sediments containing cultivable fungi *(Acremonium fusidiodes, Penicillium alli-sativi, Penicillium chrysogenum, Penicillium palitans, Penicillium solitum, Pseudogymnoascus verrucocus)* were analysed for their trypanocidal, antimalarial, nemoticidal and herbicidal, leishmanicidal, and antifungal properties (Ogaki *et al.*, 2020). Actinobacteria isolations of 72 psychro-tolerant strains from endemic Antarctic plants revealed bioactive compounds with future anti-cancer promise (Silva *et al.*, 2020).

8.2.8 Novel Organisms

This relatively unexplored land is species-rich despite its harsh climatic conditions. Novel organisms are still being discovered, such as a Psychrobacter species from ornithogenic soils (Adelie penguin colony) (Bowman *et al.*, 1996). Similarly, genetically distinct *Pseudoalteromonas antarctica* sp. from coastal environment showed phenotypic similarities with known species *Pseudoalteromonas haloplanktis* and *Pseudoalteromonas atlantica* (Bozal *et al.*, 1997). Two strains from Antarctic sea ice were shown to produce omega-3 eisopentaenoic acid (E.P.A.) (20:5w3), ranging from ~2-16% of total fatty acids composition, were denoted as *Shewanella frigidimarina* sp. and *Shewanella gelidimarina* sp. respectively (Bowman *et al.*, 1997). A novel *Flavobacterium* sp. was also isolated from an Adelie penguin colony on St. King George's Island, Antarctica (Yi *et al.*, 2005).

8.2.9 Human Impact

The detrimental effects of human pressures at Arctowski Station on King George's Island, Antarctica, condoned the introduction of alien species trailed by breeding sites disturbances to birds and seals and their migration as studied by Chwedorzewska and Korczak (2010). A risk assessment was conducted by Chown *et al.* (2012) to quantify the likelihood of establishing introduced species in the Antarctic.

8.3 Marine and Freshwater Environments (National Perspectives)

Considering the late admittance into the Antarctic Treaty System, Indian scientists excelled in research from 1981-1982.

8.3.1 Marine environments

The study conducted by Kureishy *et al.* (1993) projected heavy metals estimations of krill and zooplankton populations in the Southern Ocean had shown lead (0.05 to 8.0 μg g-1) and copper (6 to 160 μg g-1) ranges.

8.3.2 Freshwater environments

A study conducted by Tiwari and Kulkarni (2006) found that the glacial lakes (West and North West of Maitri station) divulge higher Fe concentrations concerning other lakes in the study. Priyadarshini lake (Zub lake close to Maitri station) was also analysed for heavy metal concentrations in the ranges of Fe- 0.191 to 1.655 mg/L; Mn- 0.005 to 0.39 mg/L, and Pb- B.D.L. to 0.348 mg/L. Shannon weaver indices calculated for phytoplankton diversity indicated low productivity for Priyadarshini lake and glacial lakes sampled closed to Maitri station (Tiwari, 2006).

Sediment core lithology from Antarctic lakes showed natural sources of rock weathering and biogenic algal processes (Choudhary *et al.*, 2018). The dynamism of deglaciated lacustrine systems was further assessed by rock interaction associations and sea spray influences (Magesh *et al.*, 2020).

8.3.3 Wastewater environments

Rotating Biological Contactor (R.B.C.) installed at Maitri station affirmed inlet of kitchen wastewater contained high quantities of oil and grease, suspended solids, heavy metals, C.O.D., and B.O.D on treatment reduced drastically (Tiwari, 2006). Efforts have been made to increase R.B.C.s efficacy by prolonging solids' settling time. There were elevated levels of oil, grease, and hydrocarbons present in the soil near the Bhaskara generator at Maitri station (Tiwari, 2006) and described an environmental management plan to formulate environment impact assessments and enforce corrective actions. In the same year, Tiwari *et al.* (2006b) studied the efficiency of the rotating biological contactors (R.B.C.s) installed at the Maitri station for the effluent treatment.

Baseline studies detected a diversity of ammonium oxidizers in the McMurdo Dry Valley ecosystem (Shravage *et al.*, 2007). Ghosh and Ramteke (2008) stipulated procedural guidelines to manage the air, soil, and water quality around Maitri station in 2008.

8.3.4 Radiation

Uranium (U238) contents in soil (Maitri station) ranged from 0.036 to 0.364 ppm, lower than the soils found in India (Siingh *et al.*, 2013). Radiation exhalation rates in Bharati station (222Rn/220Rn) were double that of Maitri station due to the topography but not a threat to temporary station members (Prajith *et al.*, 2019).

8.3.5 Air environments

Photochemical conversions by dimethyl sulfoxide (D.M.S.) in the oceans were determining factors affecting aerosol size distribution around Maitri station as indicated by Deshpande and Kamra (2001). Scientists have observed the effects of circumpolar cyclonic storms on the aerosol particle size at Maitri Station, increasing coarse particles. PM10 of Maitri station was lower than that of McMurdo station (Gadhavi and Jayaraman, 2011). Subsequently, the size distribution of air profiles at Maitri station in normal conditions revealed a higher concentration of fine particles (Pant *et al.*, 2011). The mean aerosol optical depth measured was similar at Maitri, and Larsemann Hills with an estimated αAOD spectral dominance over Maitri curbed by a blizzard event (Chaubey *et al.*, 2011). PM10 and PM2.5 concentrations across Antarctic coastlines were 5.13 µg m-3 and 4.3 µg m-3, respectively, with coarse dominating over fine particles (Budhavant *et al.*, 2015). The total suspended particulate matter characterization disclosed Maitri- 9 µg/m3 and Bharati- 7.6 µg/m3 concentrates, respectively (Ali *et al.*, 2015). The total aerosol mass over the Maitri station has increased with time (Ranjan *et al.*, 2015). Mean mass concentrations were also calculated for Bharati station as PM1- 0.6 µg m-3, PM2.5-1 µg m-3, and PM10- 1.5 µg m-3 while over Maitri concentrations were 2.5 times more (Budhavant *et al.*, 2017). The measurement of an aerosol optical thickness (>the highest solar zenith angle) was executed over 24-hour recordings at Bharati station (Taori *et al.*, 2018).

The baselines of carbon dioxide and methane over Bharati station in Antarctica showed a gradual annual increase (Jain *et al.*, 2005). The ambient air quality at Maitri station revealed SPM-6.0 to 69.7 µg/m^3, SO2- B.D.L. to 10.1 µg/m^3, and NOx- B.D.L. to 21.1 µg/m3 measured by (Tiwari and Kulkarni, 2006).

During the period 2008-2009, black carbon values were ten times more than Bharati (Babu *et al.*, 2019). In 2010, black carbon aerosol concentration was assessed over Maitri, and Larsemann Hills showed B.C. levels of Maitri > L.H. due to humans (Chaubey *et al.*, 2010). The scavenging effect of snow for aerosols and atmospheric ions was increased by high wind speed drift (Kamra *et al.*, 2009). The scavenging ratio of snow concerning black carbon in the Arctic and Antarctic was chiefly capricious (Gogoi *et al.*, 2018). The presence of greenhouse gas (CO_2) at Bharati Station, Larsemann Hills, was observed when favourable climatic conditions drew in the gas from the oceanic region (Pathakoti *et al.*, 2018). Ramesh and Soni (2018) presented meteorological data under the READER project dictating minor statistical climate-change over the last quarter. Monitoring of ozone holes during spring via ozonesonde satellite transmissions revealed profound changes over the Bharati station (Hulswar *et al.*, 2020).

8.3.6 Bio-prospecting advances

Marine bacteria are often selected for bioremediation projects because of their innate ability to withstand fluctuations in pH, salinity, temperature, currents, sea surface temperature, and precipitation, along with nutrient concentrations. Bacterial isolates from Antarctic marine waters have shown varying degrees of resistance to chloramphenicol, tetracycline, ampicillin, streptomycin, and kanamycin with heavy metals like cadmium, chromium, zinc, and mercury (De Souza *et al.*, 2006). Bio-surfactant properties were found in isolates from Antarctic seawater (*Oceanobacillus* sp.) (Jadhav *et al.*, 2013). The reductive effect of ultraviolet B (UV-B) rays on the carotenoid content in lichen *Dermatocarpon* sp. and *Acarospora gwynii* was uncovered (Gautam *et al.*, 2011).

To address the cancer crisis in India, (Ashok *et al.*, 2019) isolated microbial L-asparaginase (glutaminase and urease free) from Antarctic soil and Moss, showing promise. As a defence mechanism, some Antarctic bacteria produced gold nanoparticles that targeted and killed sulfate-reducing bacteria (Das *et al.*, 2020).

REFERENCES

1. Abollino, O., Aceto, M., Buoso, S., Gasparon, M., Green, W.J., Malandrino, M., and Mentasti, E. (2004) Distribution of major, minor and trace elements in lake environments of Antarctica. Antarctic Science 16: 277.
2. Abram, N.J., Mulvaney, R., Wolff, E.W., Triest, J., Kipfstuhl, S., Trusel, L.D. *et al.* (2013) Acceleration of snowmelt in an Antarctic Peninsula ice core during the twentieth century. Nature Geoscience 6: 404-411.
3. Aislabie, J.M., Balks, M.R., Foght, J.M., and Waterhouse, E.J. (2004). Hydrocarbon spills on Antarctic soils: effects and management. Environmental science & technology 38: 1265-1274.
4. Alam, S., Singh, L., Dube, S., Reddy, G., and Shivaji, S. (2003) Psychrophilic Planococcus maitriensis sp. nov. from Antarctica. Systematic and applied microbiology 26: 505-510.
5. Ali, K., Trivedi, D., and Sahu, S. (2015) Physico-chemical characterization of total suspended particulate matter over two coastal stations of Antarctica and adjoining ocean. Atmospheric Environment 122: 531-540.
6. Ansell, D., Dicks, B., Guenette, C., Moller, T., Santner, R., and White, I. (2001). A review of the problems posed by spills of heavy fuel oils. In International Oil Spill Conference: American Petroleum Institute, pp. 591-596.
7. Antonello, A. (2014) Nature conservation and Antarctic diplomacy, 1959–1964. The Polar Journal 4: 335-353.

8. Apama, N., Ramani, A., and Nagarajac, R. (2014) Risk management support through India Remote Sensing Satellites. International Archives of the Photogrammetry, Remote Sensing & Spatial Information Sciences.

9. Areas, A.S.P., Form, P.A., and Treaty, A. (2001) Antarctic Conservation Act of 1978. Public Law 95: 541.

10. Ashok, A., Doriya, K., Rao, J.V., Qureshi, A., Tiwari, A.K., and Kumar, D.S. (2019). Microbes producing L-asparaginase free of glutaminase and urease isolated from extreme locations of Antarctic soil and Moss. Scientific reports 9: 1-10.

11. Babu, S.S., Chaubey, J.P., Nair, V.S., and Moorthy, K.K. (2019) Characterization of Atmospheric aerosols over Indian Antarctic station during the Southern Hemispheric summer of 2008-2009.

12. Bansal, P., Sannd, R., Srikanth, N., and Lavekar, G. (2013) Effect of Certain established Ayurvedic Rasayana Food Supplements on Aging Process at Antarctica-A review.

13. Bargagli, R. (2006) Antarctic ecosystems: environmental contamination, climate change, and human impact: Springer Science & Business Media.

14. Bargagli, R. (2008) Environmental contamination in Antarctic ecosystems. Science of the total environment 400: 212-226.

15. Bargagli, R., Monaci, F., and Bucci, C. (2007) Environmental biogeochemistry of mercury in Antarctic ecosystems. Soil Biology and Biochemistry 39: 352-360.

16. Barnes, D.K., and Conlan, K.E. (2007) Disturbance, colonization and development of Antarctic benthic communities. Philosophical Transactions of the Royal Society B: Biological Sciences 362: 11-38.

17. Barnes, D.K., Walters, A., and Goncalves, L. (2010) Macroplastics at sea around Antarctica. Mar Environ Res 70: 250-252.

18. Barr, S. (2018) Twenty years of protection of historic values in Antarctica under the Madrid Protocol. The Polar Journal 8: 241-264.

19. Bastmeijer, K. (2009) A long term strategy for Antarctic tourism: the key to decision making within the Antarctic treaty system? POLAR TOURISM: HUMAN, ENVIRONMENTAL AND GOVERNANCE DIMENSIONS, Cognizant Communication Corporation, Patrick Maher, Emma Stewart and Michael Lück, eds, Elmsford, NY.

20. Bastmeijer, K., and Roura, R. (2004) Regulating Antarctic tourism and the precautionary principle. The American Journal of International Law 98: 763-781.

21. Bastmeijer, K., and Roura, R. (2008) Environmental impact assessment in Antarctica. In Theory and practice of transboundary environmental impact assessment: Brill Nijhoff, pp. 175-219.

22. Bauer, T. (2013) Tourism in the Antarctic: opportunities, constraints, and future prospects: Routledge.

23. Bauer, T., and Dowling, R.K. (2006) 18 The Antarctic Cruise Industry. Cruise ship tourism: 195.

24. Baust, J.G., and Lee Jr, R.E. (1983) Population differences in antifreeze/cryoprotectant accumulation patterns in an Antarctic insect. Oikos: 120-124.

25. Beck, P.J. (1989) Convention on the Regulation of Antarctic Mineral Resource Activities: a major addition to the Antarctic Treaty system. Polar Record 25: 19-32.

26. Beck, P.J. (1994) Managing Antarctic tourism: a front-burner issue. Annals of Tourism Research 21: 375-386.

27. Behrendt, J.C. (1983) Petroleum and mineral resources of Antarctica. In: Geological Survey, Washington, DC (USA).

28. Behrendt, J.C. (2005) The Ninth Circle: A Memoir of Life and Death in Antarctica, 1960-1962: UNM Press.

29. Belova, E., Voelger, P., Kirkwood, S., Hagelin, S., Lindskog, M., Körnich, H. *et al.* (2020) Validation of wind measurements of two MST radars in northern Sweden and in Antarctica. Atmospheric Measurement Techniques Discussions: 1-17.

30. Bentley, C.R. (2014) Antarctica: A Biography. Arctic, Antarctic, and Alpine Research 46: 528-531, 524.

31. Berkman, P., and Young, O. (2005) Antarctic Treaty Summit: Washington, DC (2009). In AGU Fall Meeting Abstracts, pp. ED11D-1124.

32. Bester, M., Ferguson, J., and Jonker, F. (2002) Population densities of pack-ice seals in the Lazarev Sea, Antarctica. Antarctic Science 14: 123-127.

33. Beyer, L., and Bölter, M. (2000) Chemical and biological properties, formation, occurrence and classification of Spodic Cryosols in a terrestrial ecosystem of East Antarctica (Wilkes Land). Catena 39: 95-119.

34. Bharti, P.K., Sharma, B., Singh, R., and Tyagi, A. (2016) Waste generation and management in Antarctica. Procedia Environmental Sciences 35: 40-50.

35. Blackie, A. (2016) Sovereignty on Ice: The Status of Antarctica in International Law. UNSWLJ Student Series.

36. Blay, S., and Green, J. (1995) The development of a liability annex to the Madrid Protocol. Envtl Poly & L 25: 24.

37. Blay, S.K. (1992) New trends in the protection of the Antarctic environment: the 1991 Madrid Protocol. The American Journal of International Law 86: 377-399.

38. Boger, S.D. (2011) Antarctica—before and after Gondwana. Gondwana Research 19: 335-371.

39. Bonacci, S., Browne, M.A., Dissanayake, A., Hagger, J.A., Corsi, I., Focardi, S., and Galloway, T.S. (2004) Esterase activities in the bivalve mollusc Adamussium colbecki as a biomarker for pollution monitoring in the Antarctic marine environment. Marine pollution bulletin 49: 445-455.

40. Bou Franch, V. (1996) Waste Disposal and Waste Management in Antarctica and the Southern Ocean.

41. Bowman, J.P., Cavanagh, J., Austin, J.J., and Sanderson, K. (1996) Novel Psychrobacter species from Antarctic ornithogenic soils. International journal of systematic and evolutionary microbiology 46: 841-848.

42. Bowman, J.P., McCammon, S.A., Brown, M.V., Nichols, D.S., and McMeekin, T.A. (1997) Diversity and association of

psychrophilic bacteria in Antarctic sea ice. Applied and Environmental Microbiology 63: 3068-3078.

43. Bozal, N., Tudela, E., Rossello-Mora, R., Lalucat, J., and Guinea, J. (1997) Pseudoalteromonas antarctica sp. nov., isolated from an Antarctic coastal environment. International Journal of Systematic and Evolutionary Microbiology 47: 345-351.

44. Broady, P.A. (1989) Broadscale patterns in the distribution of aquatic and terrestrial vegetation at three ice-free regions on Ross Island, Antarctica. In: High Latitude Limnology: Springer, pp. 77-95.

45. Brockett, D. (2005) Antarctic Katabatic Winds.

46. Bruni, V., Maugeri, T., and Monticelli, L. (1997) Faecal pollution indicators in the Terra Nova Bay (Ross Sea, Antarctica). Marine Pollution Bulletin 34: 908912.

47. Buck, E.H. (2010) Ballast water management to combat invasive species.

48. Budhavant, K., Safai, P., and Rao, P. (2015) Sources and elemental composition of summer aerosols in the Larsemann Hills (Antarctica). Environmental Science and Pollution Research 22: 2041-2050.

49. Budhavant, K., Rao, P., and Safai, P. (2017) Size distribution and chemical composition of summer aerosols over Southern Ocean and the Antarctic region. Journal of Atmospheric Chemistry 74: 491-503.

50. Bueno, C., Kandratavicius, N., Venturini, N., Figueira, R., Pérez, L., Iglesias, K., and Brugnoli, E. (2018) An Evaluation of Trace Metal Concentration in Terrestrial and Aquatic Environments near Artigas Antarctic Scientific Base (King George Island, Maritime Antarctica). Water, Air, & Soil Pollution 229: 398.

51. Bull, C. (1966) Climatological observations in ice-free areas of Southern Victoria Land, Antarctica. Studies in Antarctic Meteorology 9: 177-194.

52. Burton-Johnson, A., Black, M., Fretwell, P.T., and Kaluza-Gilbert, J. (2016) An automated methodology for differentiating rock from snow, clouds and sea in Antarctica from Landsat 8

imagery: a new rock outcrop map and area estimation for the entire Antarctic continent. The Cryosphere 10: 1665-1677.

53. Butt, N. (2007) The impact of cruise ship generated waste on home ports and ports of call: A study of Southampton. Marine Policy 31: 591-598.

54. Cabrerizo, A., Dachs, J., Barceló, D., and Jones, K.C. (2013) Climatic and biogeochemical controls on the remobilization and reservoirs of persistent organic pollutants in Antarctica. Environmental science & technology 47: 42994306.

55. Calder-Steele, N., Hogarth, K., McArthur, N., and McTurk, L. (2013) Consultative party status and alternative governance systems in the Antarctic.

56. Cardone, I.J. (2019) A continent for peace and science: Antarctic science and international politics from the 6th International Geographical Congress to the Antarctic Treaty (1895-1959). In: Universidade de São Paulo.

57. Carrasco, J.F., Bromwich, D.H., and Liu, Z. (1997) Mesoscale cyclone activity over Antarctica during 1991: 1. Marie Byrd Land. Journal of Geophysical Research: Atmospheres 102: 13923-13937.

58. Carter, L., McCave, I., and Williams, M.J. (2008) Circulation and water masses of the Southern Ocean: a review. Developments in earth and environmental sciences 8: 85-114.

59. Caruso, G. (2019) Microplastics as vectors of contaminants. Marine pollution bulletin 146: 921-924.

60. Casal, P., Casas, G., Vila-Costa, M., Cabrerizo, A., Pizarro, M., Jiménez, B., and Dachs, J. (2019) Snow amplification of persistent organic pollutants at coastal Antarctica. Environmental science & technology 53: 8872-8882.

61. Chaubey, J.P., Moorthy, K.K., Babu, S.S., and Nair, V.S. (2011) The optical and physical properties of atmospheric aerosols over the Indian Antarctic stations during southern hemispheric summer of the International Polar Year 2007–2008. Ann Geophys 29: 109-121.

62. Chaubey, J.P., Moorthy, K.K., Babu, S.S., Nair, V.S., and Tiwari, A. (2010) Black carbon aerosols over coastal Antarctica and its

scavenging by snow during the Southern Hemispheric summer. Journal of Geophysical Research: Atmospheres 115.

63. Chen, L. (2013) Antarctic Governance: From ATCM to A Permanent Antarctic Organization?

64. Chenery, H.B. (1960) Patterns of industrial growth. The American economic review 50: 624-654.

65. Choudhary, S., Nayak, G., Tiwari, A.K., and Khare, N. (2018) Sediment composition and its effect on the productivity in Larsemann Hills, East Antarctica. Arabian Journal of Geosciences 11: 416.

66. Chow, J.C., and Watson, J.G. (2002) Review of PM2. 5 and PM10 apportionment for fossil fuel combustion and other sources by the chemical mass balance receptor model. Energy & Fuels 16: 222-260.

67. Chown, S.L., Huiskes, A.H., Gremmen, N.J., Lee, J.E., Terauds, A., Crosbie, K. *et al.* (2012) Continent-wide risk assessment for the establishment of nonindigenous species in Antarctica. Proceedings of the National Academy of Sciences 109: 4938-4943.

68. Chown, S.L., Brooks, C.M., Terauds, A., Le Bohec, C., van KlaverenImpagliazzo, C., Whittington, J.D. *et al.* (2017) Antarctica and the strategic plan for biodiversity. PLoS biology 15: e2001656.

69. Chu, W.-L., Dang, N.-L., Kok, Y.-Y., Yap, K.-S.I., Phang, S.-M., and Convey, P. (2019) Heavy metal pollution in Antarctica and its potential impacts on algae. Polar Science 20: 75-83.

70. Chwedorzewska, K.J., and Korczak, M. (2010) Human impact upon the environment in the vicinity of Arctowski Station, King George Island, Antarctica. Polish Polar Research: 45-60.

71. Clarkson, P. (1997) Antarctica: Wilderness or Wasteland? ALPINE JOURNAL 102: 8-12.

72. Clucas, G.V., Dunn, M.J., Dyke, G., Emslie, S.D., Levy, H., Naveen, R. *et al.* (2014) A reversal of fortunes: climate change 'winners' and 'losers' in Antarctic Peninsula penguins. Scientific reports 4: 1-7.

73. Coetzee, B.W., Convey, P., and Chown, S.L. (2017) Expanding the protected area network in Antarctica is urgent and readily achievable. Conservation Letters 10: 670-680.

74. Collins, M.A., Brickle, P., Brown, J., and Belchier, M. (2010) The Patagonian toothfish: biology, ecology and fishery. Advances in marine biology 58: 227-300.

75. COMNAP (2005) Practical Guidelines for Developing and Designing Environmental Monitoring Programmes in Antarctica.

76. Congress, U. (1989) Polar Prospects: A Minerals Treaty for Antarctica. Office of Technology Assessment, US Government Printing Office, Washington, DC URL https://www princeton edu/~ ota/disk1/1989/8926/8926 PDF.

77. Constable, A.J., de la Mare, W.K., Agnew, D.J., Everson, I., and Miller, D. (2000) Managing fisheries to conserve the Antarctic marine ecosystem: practical implementation of the Convention on the Conservation of Antarctic Marine Living Resources (CCAMLR). ICES Journal of Marine Science 57: 778-791.

78. Convey, P. (2010) Terrestrial biodiversity in Antarctica–Recent advances and future challenges. Polar Science 4: 135-147.

79. Convey, P., and Peck, L.S. (2019) Antarctic environmental change and biological responses. Science Advances 5: eaaz0888.

80. Convey, P., Hughes, K.A., and Tin, T. (2012) Continental governance and environmental management mechanisms under the Antarctic Treaty System: sufficient for the biodiversity challenges of this century? Biodiversity 13: 234248.

81. Corsolini, S., Baroni, D., Martellini, T., Pala, N., and Cincinelli, A. (2019) PBDEs and PCBs in terrestrial ecosystems of the Victoria Land, Antarctica. Chemosphere 231: 233-239.

82. Council, N.R., and Committee, M.S. (2007) Successful response starts with a map: improving geospatial support for disaster management: National Academies Press.

83. Craig, R.K. (2012) Comparative ocean governance: place-based protections in an era of climate change: Edward Elgar Publishing.

84. Cullen, R. (1994) Antarctic minerals and conservation. Ecological Economics 10: 143-155.

85. Curl, H.C., Barton, K., and Harris, L. (1992) Oil spill case histories, 19671991: summaries of significant US and international spills.

86. Das, K.R., Tiwari, A.K., and Kerkar, S. (2020) Psychrotolerant Antarctic bacteria biosynthesize gold nanoparticles active against sulphate reducing bacteria. Preparative Biochemistry & Biotechnology 50: 438-444.

87. Dasgupta, S., and Sengupta, P. (2003) Granulite Belt, India. Proterozoic East Gondwana: Supercontinent Assembly and Breakup 206: 131.

88. De Lange, H., Sala, S., Vighi, M., and Faber, J. (2010) Ecological vulnerability in risk assessment—a review and perspectives. Science of the total environment 408: 3871-3879.

89. De Souza, M.-J., Nair, S., Bharathi, P.L., and Chandramohan, D. (2006) Metal and antibiotic-resistance in psychrotrophic bacteria from Antarctic Marine waters. Ecotoxicology 15: 379-384.

90. Deegan, C., and Gordon, B. (1996) A study of the environmental disclosure practices of Australian corporations. Accounting and business research 26: 187199.

91. Delille, D., Basseres, A., and Dessommes, A. (1998) Effectiveness of bioremediation for oil-polluted Antarctic seawater. Polar Biology 19: 237-241.

92. Derraik, J.G. (2002) The pollution of the marine environment by plastic debris: a review. Marine pollution bulletin 44: 842-852.

93. Deshpande, C., and Kamra, A. (2001) Diurnal variations of the atmospheric electric field and conductivity at Maitri, Antarctica. Journal of Geophysical Research: Atmospheres 106: 14207-14218.

94. Dodds, K. (2010) Governing Antarctica: contemporary challenges and the enduring legacy of the 1959 Antarctic Treaty. Global Policy 1: 108-115.

95. Dodds, K., and Hemmings, A.D. (2013) Britain and the British Antarctic Territory in the wider geopolitics of the Antarctic and the Southern Ocean. International Affairs 89: 1429-1444.

96. Dowling, R.K. (2006) Cruise ship tourism: CABI.

97. Duarte, A.W.F., dos Santos, J.A., Vianna, M.V., Vieira, J.M.F., Mallagutti, V.H., Inforsato, F.J. *et al.* (2018) Cold-adapted enzymes produced by fungi from terrestrial and marine Antarctic environments. Critical reviews in biotechnology 38: 600-619.

98. Eerkes-Medrano, D., Thompson, R.C., and Aldridge, D.C. (2015) Microplastics in freshwater systems: a review of the emerging threats, identification of knowledge gaps and prioritisation of research needs. Water Research 75: 63-82.

99. Eklund, C.R. (1964) The Antarctic skua. Scientific American 210: 94-102.

100. Ellenberg, U. (2017) Impacts of penguin tourism. In: Ecotourism's Promise and Peril: Springer, pp. 117-132.

101. Esteban, S., Moreno-Merino, L., Matellanes, R., Catalá, M., Gorga, M., Petrovic, M. *et al.* (2016) Presence of endocrine disruptors in freshwater in the northern Antarctic Peninsula region. Environmental Research 147: 179-192.

102. Evans, S.l. (2011) Icy heritage–managing historic sites in the Antarctic: Pristine wilderness, anthropogenic degradation or cultural landscape? The Polar Journal 1: 87-100.

103. Fabra, A., and Gascón, V. (2008) The Convention on the Conservation of Antarctic Marine Living Resources (CCAMLR) and the ecosystem approach. The International Journal of Marine and Coastal Law 23: 567-598.

104. Feller, G., Thiry, M., and Gerday, C. (1990) Sequence of a lipase gene from the antarctic psychrotroph Moraxella TA144. Nucleic acids research 18: 6431.

105. Ferrada, L.V. (2018) Five factors that will decide the future of Antarctica. The Polar Journal 8: 84-109.

106. Ferreira, M.F., Nostro, F.L., Honji, R., Ansaldo, M., and Genovese, G. (2019) Endocrine and reproductive endpoints of Notothenia rossii and N. coriiceps: A baseline study for

ecotoxicological monitoring in Antarctic waters. Marine pollution bulletin 145: 418-428.

107. Floren, D.W. (2001) Antarctic mining regimes: an appreciation of the attainable. J Envtl L & Litig 16: 467.

108. Fogarty, E. (2011) Antarctica: assessing and protecting Australia's national interests: JSTOR.

109. Franklin, C.M. (1961) Appendix L: Twelve-State Treaty Guaranteeing Nonmilitarization of Antarctica and Freedom Of Scientific Investigation. International Law Studies 53: 3.

110. French, D. (1999) Sustainable development and the 1991 Madrid protocol to the 1959 Antarctic treaty: The primacy of protection in a particularly sensitive environment. Journal of International Wildlife Law and Policy 2: 291-317.

111. Fretwell, P., Pritchard, H.D., Vaughan, D.G., Bamber, J.L., Barrand, N.E., Bell, R. *et al.* (2013) Bedmap2: improved ice bed, surface and thickness datasets for Antarctica. The Cryosphere 7: 375-393.

112. Friedmann, E.I., and Ocampo, R. (1976) Endolithic blue-green algae in the dry valleys: primary producers in the Antarctic desert ecosystem. Science 193: 1247-1249.

113. Gadhavi, H., and Jayaraman, A. (2011) Aerosal Characteristics over Maitri Antarctica and Their Impact on Radiative Forcing.

114. Gandhi, N., Ramesh, R., Laskar, A., Sheshshayee, M., Shetye, S., Anilkumar, N. *et al.* (2012) Zonal variability in primary production and nitrogen uptake rates in the southwestern Indian Ocean and the Southern Ocean. Deep Sea Research Part I: Oceanographic Research Papers 67: 32-43.

115. Gasparon, M., and Burgess, J. (2000) Human impacts in Antarctica: traceelement geochemistry of freshwater lakes in the Larsemann Hills, East Antarctica. Environmental Geology 39: 963-976.

116. Gautam, S., Singh, J., and Pant, A.B. (2011) Effect of UV-B radiations on the pigments of two Antarctic lichens of Schirmacher Oasis, East Antarctica. Polish Polar Research: 279-287.

117. Ghosh, T., and Ramteke, D. (2008) Environmental Status at Maitri, Indian Permanent Station in Antarctica: Preliminary Studies.

118. Gilpin, A., and West, A. (1995) Environmental impact assessment: cutting edge for the 21st century: Cambridge University Press.

119. Giudice, A.L., Caruso, G., Rizzo, C., Papale, M., and Azzaro, M. (2019) Bacterial communities versus anthropogenic disturbances in the Antarctic coastal marine environment. Environmental Sustainability: 1-14.

120. Glatstein, D.A., Bruna, N., Gallardo-Benavente, C., Bravo, D., Carro Perez, M.E., Francisca, F.M., and Pérez-Donoso, J.M. (2018) Arsenic and cadmium bioremediation by Antarctic bacteria capable of biosynthesizing CdS fluorescent nanoparticles. Journal of Environmental Engineering 144: 04017107.

121. Gogoi, M.M., Babu, S.S., Pandey, S.K., Nair, V.S., Vaishya, A., Girach, I., and Koushik, N. (2018) Scavenging ratio of black carbon in the Arctic and the Antarctic. Polar Science 16: 10-22.

122. Goldsworthy, L., and Hemmings, A.D. (2009) The Antarctic protected area approach. Shared resources: Issues of governance: 105-128.

123. Golubev, S. (2020) Macroplastic in Seabirds at Mirny, Antarctica. Birds 1: 13-18.

124. González-Pleiter, M., Edo, C., Velázquez, D., Casero-Chamorro, M.C., Leganés, F., Quesada, A. *et al.* (2020) First detection of microplastics in the freshwater of an Antarctic Specially Protected Area. Marine Pollution Bulletin. 161: 111811.

125. Goudie, A.S. (2018) Human impact on the natural environment: John Wiley & Sons.

126. Gries, C. (1996) Lichens as indicators of air pollution. Lichen biology 63.

127. Gröndahl, F., Sidenmark, J., and Thomsen1, A. (2009) Survey of wastewater disposal practices at Antarctic research stations. Polar Research 28: 298-306.

128. Gupta, U.K., Joshi, P., Bodani, P., Oza, S.R., Rajak, D.R., and Oza, M. (2019) Development of a Web-GIS based system for safer ship navigation in Antarctic region using open source technologies. Journal of Geomatics 13.

129. Gutt, J., Hosie, G., and Stoddart, M. (2010) Marine life in the Antarctic. Life in the World's Oceans: 203-220.

130. Haase, D., Lamers, M., and Amelung, B. (2009) Heading into uncharted territory? Exploring the institutional robustness of self-regulation in the Antarctic tourism sector. Journal of Sustainable Tourism 17: 411-430.

131. Halverson, S.M. (2004) Small State with a Big Tradition: Norway Continues Whaling at the Expense for Integration and Nordic Cooperation. Syracuse J Int'l L & Com 31: 121.

132. Hampson, E. (2001) Antarctic ship-borne tourism–an analysis of the 'Measure of Compliance'of approvals, management plans, tour operator's EIA, New Zealand legislation & the guidelines pertaining to tourists visiting the Ross Sea region, Antarctica.

133. Hanevold, T. (1971) The Antarctic Treaty Consultative Meetings —Form and Procedure. Cooperation and Conflict 6: 183-199.

134. Hansen, A.D., Lowenthal, D.H., Chow, J.C., and Watson, J.G. (2001) Black carbon aerosol at McMurdo Station, Antarctica. Journal of the Air & Waste Management Association 51: 593-600.

135. Harris, C.M. (2002) Standardisation of zones within specially protected and managed areas under the Madrid Protocol. Unpublished) Colin Harris 5.

136. Harris, C.M., and Meadows, J. (1992) Environmental management in Antarctica Instruments and institutions. Marine Pollution Bulletin 25: 239-249.

137. Hayashi, M. (1986) The Antarctica Question in the United Nations. Cornell Int'l LJ 19: 275.

138. Hayton, R.D. (1960) The Antarctic settlement of 1959. Am J Int'l L 54: 349.

139. Hodgson-Johnston, I., Jackson, A., Jabour, J., and Press, A. (2017) Cleaning up after human activity in Antarctica: legal obligations and remediation realities. Restoration Ecology 25: 135-139.

140. Hodson, A., Nowak, A., Sabacka, M., Jungblut, A., Navarro, F., Pearce, D. *et al.* (2017) Climatically sensitive transfer of iron to maritime Antarctic ecosystems by surface runoff. Nature Communications 8: 1-7.

141. Hughes, K.A. (2004) Reducing sewage pollution in the Antarctic marine environment using a sewage treatment plant. Marine Pollution Bulletin 49: 850853.

142. Hughes, K.A., and Convey, P. (2010) The protection of Antarctic terrestrial ecosystems from inter-and intra-continental transfer of non-indigenous species by human activities: a review of current systems and practices. Global Environmental Change 20: 96-112.

143. Hughes, K.A., and Convey, P. (2014) Alien invasions in Antarctica—is anyone liable? Polar Research 33: 22103.

144. Hulme, P.E. (2006) Beyond control: wider implications for the management of biological invasions. Journal of Applied Ecology 43: 835-847.

145. Hulswar, S., Soni, V., Satpate, J., More, R., and Mahajan, A.S. (2020) Validation of satellite retrieved ozone profiles using in-situ ozonesonde observations over the Indian Antarctic station, Bharati. Polar Science: 100547.

146. Hunt Jr, G.L., Heinemann, D., and Everson, I. (1992) Distributions and predator-prey interactions of macaroni penguins, Antarctic fur seals, and Antarctic krill near Bird Island, South Georgia. Marine Ecology Progress Series: 15-30.

147. Huston, S.M. (2001) Gone, but maybe not forever... The huskies in Antarctica.

148. Ivanov, V. (1985) Prediction of the oil and gas potential of Antarctica on the basis of geological conditions. Polar Geography 9: 116-131.

149. Jabour, J. (2008) Successful Conservation—Then What? The De-Listing of Arctocephalus Fur Seal Species in Antarctica. Journal of International Wildlife Law & Policy 11: 1-29.

150. Jackson, T.A. (1997) Long-range atmospheric transport of mercury to ecosystems, and the importance of anthropogenic emissions A critical review and evaluation of the published evidence. Environmental Reviews 5: 99-120.

151. Jadhav, V.V., Yadav, A., Shouche, Y.S., Aphale, S., Moghe, A., Pillai, S. *et al.* (2013) Studies on biosurfactant from Oceanobacillus sp. BRI 10 isolated from Antarctic seawater. Desalination 318: 64-71.

152. Jain, S., Ghude, S.D., Kumar, A., Arya, B., and Kulkarni, P. (2005) Continuous observations of surface air concentration of carbon dioxide and methane at Maitri, Antarctica. Current Science: 1941-1948.

153. Jawak, S.D., and Luis, A.J. (2013) Very high-resolution satellite data for improved land cover extraction of Larsemann Hills, Eastern Antarctica. Journal of Applied Remote Sensing 7: 073460.

154. Jawak, S.D., Udhayaraj, A., and Luis, A.J. (2016a) Geospatial mapping of vegetation in the Antarctic environment using very high-resolution World View 2 imagery. In: Land Surface and Cryosphere Remote Sensing III: International Society for Optics and Photonics, p. 98772N.

155. Jawak, S.D., Jadhav, A., and Luis, A.J. (2016b) Object-oriented feature extraction approach for mapping supraglacial debris in Schirmacher Oasis using very high-resolution satellite data. Inland surface and cryosphere remote sensing III: International Society for Optics and Photonics, p. 98772L.

156. Jawak, S.D., Panditrao, S.N., and Luis, A.J. (2016c) C-band RISAT-1 imagery for geospatial mapping of cryospheric surface features in the Antarctic environment. In Earth Observing Missions and Sensors: Development, Implementation, and Characterization IV: International Society for Optics and Photonics, p. 98811R.

157. Johnson, M. (2006) Liability for Environmental Damage in Antarctica: The Adoption of Annex VI to the Antarctic Environmental Protocol. Geo Int'l Envtl L Rev 19: 33.

158. Joseph, B., Ramteke, P.W., and Thomas, G. (2008) Cold active microbial lipases: some hot issues and recent developments. Biotechnology advances 26: 457-470.

159. Joyner, C.C. (1988) The evolving Antarctic minerals regime. Ocean Development & International Law 19: 73-95.

160. Jungblut, A.D., Allen, M.A., Burns, B.P., and Neilan, B.A. (2009) Lipid biomarker analysis of cyanobacteria-dominated microbial mats in meltwater ponds on the McMurdo Ice Shelf, Antarctica. Organic Geochemistry 40: 258-269.

161. Kahle, J., and Zauke, G.-P. (2003) Trace metals in Antarctic copepods from the Weddell Sea (Antarctica). Chemosphere 51: 409-417.

162. Kakareka, S., Kukharchyk, T., and Kurman, P. (2019) Major and trace elements content in freshwater lakes of Vecherny Oasis, Enderby Land, East Antarctica. Environmental Pollution 255: 113126.

163. Kamra, A., Siingh, D., and Pant, V. (2009) Scavenging of atmospheric ions and aerosols by drifting snow in Antarctica. Atmospheric research 91: 215-218.

164. Kanfoush, S.L., Hodell, D.A., Charles, C.D., Guilderson, T.P., Mortyn, P.G., and Ninnemann, U.S. (2000) Millennial-scale instability of the Antarctic ice sheet during the last glaciation. Science 288: 1815-1819.

165. Kennicutt, I., Mahlon C, McDonald, S.J., Sericano, J.L., Boothe, P., Oliver, J., Safe, S. *et al.* (1995) Human contamination of the marine environment-Arthur Harbor and McMurdo Sound, Antarctica. Environmental science & technology 29: 1279-1287.

166. Kennicutt, M.C., Chown, S.L., Cassano, J.J., Liggett, D., Massom, R., Peck, L.S. *et al.* (2014) Polar research: six priorities for Antarctic science. Nature News 512: 23.

167. Kerry, K., and Riddle, M. (2009) Health of Antarctic wildlife: an introduction. In Health of Antarctic Wildlife: Springer, pp. 1-10.

168. Keskitalo, J., Leppäranta, M., and Arvola, L. (2013) First records of primary producers of epiglacial and supraglacial lakes in western Dronning Maud Land, Antarctica. Polar Biology 36: 1441-1450.

169. Kirkham, N.R., Gjerde, K.M., and Wilson, A.M.W. (2020) DEEP-SEA mining: policy options to preserve the last frontier-lessons from Antarctica's mineral resource convention. Marine Policy 115: 103859.

170. Kitamoto, D., Ikegami, T., Suzuki, G.T., Sasaki, A., Takeyama, Y.-i., Idemoto, Y. *et al.* (2001) Microbial conversion of n-alkanes into glycolipid biosurfactants, mannosylerythritol lipids, by Pseudozyma (Candida Antarctica). Biotechnology letters 23: 1709-1714.

171. Knox, G.A. (2006) Biology of the Southern Ocean: CRC press.

172. Kock, K. (2007) Antarctic marine living resources-exploitation and its management in the Southern Ocean. Antarctic Science 19: 231.

173. Kodama, Y., Wendler, G., and Gosink, J. (1985) The effect of blowing snow on katabatic winds in Antarctica. Annals of Glaciology 6: 59-62.

174. Komárková, V., Poncet, S., and Poncet, J. (1990) Additional and revisited localities of vascular plants Deschampsia Antarctica Desv. and Colobanthus quitensis (Kunth) Bartl. in the Antarctic Peninsula area. Arctic and Alpine Research 22: 108-113.

175. Komárková, V.r., Poncet, S., and Poncet, J. (1985) Two native Antarctic vascular plants, Deschampsia Antarctica and Colobanthus quitensis: a new southernmost locality and other localities in the Antarctic Peninsula area. Arctic and Alpine Research 17: 401-416.

176. Kriwoken, L.K. (1995) Antarctic tourism and environmental impact assessment. Issues: 28.

177. Kriwoken, L.K., and Rootes, D. (2000) Tourism on ice: environmental impact assessment of Antarctic tourism. Impact assessment and project appraisal 18: 138-150.

178. Kube, M., Chernikova, T.N., Al-Ramahi, Y., Beloqui, A., Lopez-Cortez, N., Guazzaroni, M.-E. *et al.* (2013) Genome sequence and functional genomic analysis of the oil-degrading bacterium Oleispira Antarctica. Nature communications 4: 1-11.

179. Kureishy, T.W., SenGupta, R., Mesquita, A., and Sanzgiri, S. (1993) Heavy metals in some parts of Antarctica and the Southern Indian Ocean.

180. Lacerda, A., Rodrigues, L.D.S., van Sebille, E., Rodrigues, F.L., Ribeiro, L., Secchi, E.R. *et al.* (2019) Plastics in sea surface waters around the Antarctic Peninsula. Sci Rep 9: 3977.

181. Lamers, M., Stel, J., and Amelung, B. (2007) Antarctic adventure tourism and private expeditions. Prospects for polar tourism: 170-187.

182. Landan, D. (2007) Tourism in polar regions and the sub-Antarctic islands. In: Papers and Proceedings of the Royal Society of Tasmania, pp. 173-179.

183. Landau, D., and Splettstoesser, J. (2007) Management of tourism in the marine environment of Antarctica: the IAATO perspective. Tourism in Marine Environments 4: 185-193.

184. Li, H., Fu, J., Zhang, A., Zhang, Q., and Wang, Y. (2016) Occurrence, bioaccumulation and long-range transport of short-chain chlorinated paraffins on the Fildes Peninsula at King George Island, Antarctica. Environment International 94: 408-414.

185. Liu, F., Liu, J., Chen, Q., Wang, B., and Cao, Z. (2012) Pollution characteristics, ecological risk and sources of polycyclic aromatic hydrocarbons (PAHs) in surface sediment from Tuhai-Majia River system, China. Procedia Environmental Sciences 13: 1301-1314.

186. Lokabharathi, P., KrishnaKumari, L., Bhattathiri, P., and Chandramohan, D. (2001) UV Radiation and natural fluorescence linked primary production in Antarctic waters.

187. Lord, T. (2020) The Antarctic Treaty System and the peaceful governance of Antarctica: the role of the ATS in promoting peace at the margins of the world. The Polar Journal 10: 3-21.

188. Lowe, S., Browne, M., Boudjelas, S., and De Poorter, M. (2000) 100 of the world's worst invasive alien species: a selection from the global invasive species database: Invasive Species Specialist Group Auckland.

189. Ma, Y., Bian, L., Xiao, C., Allison, I., and Zhou, X. (2010) Near surface climate of the traverse route from Zhongshan Station to Dome A, East Antarctica. Antarctic Science 22: 443-459.

190. Macdonald, R.W., Mackay, D., Li, Y.-F., and Hickie, B. (2003) How will global climate change affect risks from long-range transport of persistent organic pollutants? Human and Ecological Risk Assessment 9: 643-660.

191. Magesh, N., Botsa, S.M., Dessai, S., Mestry, M., Leitao, T.D.L., and Tiwari, A. (2020) Hydrogeochemistry of the deglaciated lacustrine systems in Antarctica: Potential impact of marine aerosols and rock-water interactions. Science of The Total Environment 706: 135822.

192. Malvicini, M., Gutierrez-Moraga, A., Rodriguez, M.M., Gomez-Bustillo, S., Salazar, L., Sunkel, C. *et al.* (2018) A tricin derivative from Deschampsia antarctica Desv. inhibits colorectal carcinoma growth and liver metastasis through the induction of a specific immune response. Molecular cancer therapeutics 17: 966-976.

193. Manderson, L., and Akatsu, H. (1993) Whale meat in the diet of Ayukawa villagers. Ecology of food and nutrition 30: 207-220.

194. Martins, R.M., Nedel, F., Victoria, B.S., Guimarães, Da Silva, A.F., Colepicolo, P., De Pereira, C.M.P., and Lund, R.G. (2018) Macroalgae Extracts from Antarctica Have Antimicrobial and Anticancer Potential. Frontiers in Microbiology https://doi.org/10.3389/fmicb.2018.00412.

195. Martinez, B., Reaser, J.K., Dehgan, A., Zamft, B., Baisch, D., McCormick, C. *et al.* (2020) Technology innovation: advancing capacities for the early detection of and rapid response to invasive species. Biological Invasions 22: 75-100.

196. McColloch, R.R. (1992) Protocol on Environmental Protection to the Antarctic Treaty-The Antarctic Treaty-Antarctic Minerals

Convention-Wellington Convention-Convention on the Regulation of Antarctic Mineral Resource Activities. Ga J Int'l & Comp L 22: 211.

197. McElroy, J.K. (1984) Antarctic fisheries: history and prospects. Marine Policy 8: 239-258.

198. McGaughran, A., Terauds, A., Convey, P., and Fraser, C.I. (2019) Genomewide SNP data reveal improved evidence for Antarctic glacial refugia and dispersal of terrestrial invertebrates. Molecular Ecology 28: 4941-4957.

199. McKay, C.P., Mellon, M.T., and Friedmann, E.I. (1998) Soil temperatures and stability of ice-cemented ground in the McMurdo Dry Valleys, Antarctica. Antarctic Science 10: 31-38.

200. McKay, R., Naish, T., Carter, L., Riesselman, C., Dunbar, R., Sjunneskog, C. *et al.* (2012) Antarctic and Southern Ocean influences on Late Pliocene global cooling. Proceedings of the National Academy of Sciences 109: 6423-6428.

201. McLoughlin, S. (2001) The breakup history of Gondwana and its impact on pre-Cenozoic floristic provincialism. Australian Journal of Botany 49: 271-300.

202. McWatters, R., Wilkins, D., Spedding, T., Hince, G., Raymond, B., Lagerewskij, G. *et al.* (2016) On site remediation of a fuel spill and soil reuse in Antarctica. Science of the Total Environment 571: 963-973.

203. Messer, K., and Breth, R. (1991) Towards Firmer Institutionalization of the ATS? Future Role of the Consultative Meeting and the Issue of a Permanent Secretariat. In the Antarctic Treaty System in World Politics: Springer, pp. 379398.

204. Montarroyos, D.C.G., de Alvarez, C.E., and Bragança, L. (2019) Assessing the environmental impacts of construction in Antarctica. Environmental Impact Assessment Review 79: 106302.

205. Mosley-Thompson, E., Paskievitch, J.F., Gow, A.J., and Thompson, L.G. (1999) Late 20th century increase in South Pole snow accumulation. Journal of Geophysical Research: Atmospheres 104: 3877-3886.

206. Mulligan, J.J. (1963) Willett Range coal deposits, Mackay Glacier area, Victoria Land, Antarctica: US Department of the Interior. Bureau of Mines.

207. Nair, P., and Gupta, H. (2007) Dakshin Gangotri: The Indian Permanent Station in Antarctica.

208. Najafi, Z., Saeedi, M., Mahdavi, M., Sabourian, R., Khanavi, M., Tehrani, M.B. *et al.* (2016) Design and synthesis of novel anti-Alzheimer's agents: Acridine-chromenone and quinoline-chromenone hybrids. Bioorg Chem 67: 8494.

209. Negoita, T.G., Covaci, A., Gheorghe, A., and Schepens, P. (2003) Distribution of polychlorinated biphenyls (PCBs) and organochlorine pesticides in soils from the East Antarctic coast. Journal of Environmental Monitoring 5: 281-286.

210. Nettles, M., and Ekström, G. (2010) Glacial earthquakes in Greenland and Antarctica. Annual Review of Earth and Planetary Sciences 38.

211. Neumann, A. (2020) Wilderness Protection in Antarctica. In Wilderness Protection in Polar Regions: Brill Nijhoff, pp. 71-183.

212. Ng, C., DeMaere, M.Z., Williams, T.J., Lauro, F.M., Raftery, M., Gibson, J.A. *et al.* (2010) Metaproteogenomic analysis of a dominant green sulfur bacterium from Ace Lake, Antarctica. The ISME journal 4: 1002-1019.

213. Nichols, C.M., Lardière, S.G., Bowman, J.P., Nichols, P.D., Gibson, J.A., and Guézennec, J. (2005) Chemical characterization of exopolysaccharides from Antarctic marine bacteria. Microbial Ecology 49: 578-589.

214. Nicol, S., and Foster, J. (2016) The fishery for Antarctic krill: Its current status and management regime. In Biology and ecology of Antarctic krill: Springer, pp. 387-421.

215. O'Reilly, J., and Salazar, J.F. (2017) Inhabiting the Antarctic. The Polar Journal 7: 9-25.

216. Ogaki, M.B., Coelho, L.C., Vieira, R., Neto, A.A., Zani, C.L., Alves, T.M. *et al.* (2020) Cultivable fungi present in deep-sea sediments of Antarctica: taxonomy, diversity, and

bioprospecting of bioactive compounds. Extremophiles 24: 227238. 217. Ossenkamp, G. (2001) Antarctic Wilderness–a door to yourself?

218. Palmer, A.S., Snape, I., Stark, J.S., Johnstone, G.J., and Townsend, A.T. (2006) Baseline metal concentrations in Paramoera walkeri from East Antarctica. Marine pollution bulletin 52: 1441-1449.

219. Pandey, K., Shukla, S., Shukla, P., Giri, D., Singh, J., Singh, P., and Kashyap, A. (2004) Cyanobacteria in Antarctica: ecology, physiology and cold adaptation. CELLULAR AND MOLECULAR BIOLOGY-PARIS-WEGMANN- 50: 575-584.

220. Pandey P.C., T.A.K. (2005) Indian Contribution to Antarctic Research. In: Glimples of the Work on Environment and Development in India. Singh J.S., S.V.P. (ed). New Delhi: Angkor Publishers Pvt. Ltd., pp. 211-228.

221. Pandit, M.K., and De Wall, H. (2014) Preliminary magnetic susceptibility characteristics of the Bharati promontory (Grovness Peninsula), Larsemann Hills, Prydz Bay region, East Antarctica. Journal of the Geological Society of India 84: 163-173.

222. Pangerc, T. (2010) Baleen whale acoustic presence around South Georgia. In: University of East Anglia.

223. Pant, V., Siingh, D., and Kamra, A. (2011) Size distribution of atmospheric aerosols at Maitri, Antarctica. Atmospheric Environment 45: 5138-5149.

224. Parker, B., and Simmons, G. (1985) Paucity of nutrient cycling and absence of food chains in the unique lakes of southern Victoria Land. In Antarctic nutrient cycles and food webs: Springer, pp. 238-244.

225. Parker, K.R., and Wiens, J.A. (2005) Assessing recovery following environmental accidents: environmental variation, ecological assumptions, and strategies. Ecological Applications 15: 2037-2051.

226. Pathakoti, M., Gaddamidi, S., Gharai, B., Sudhakaran Syamala, P., Rao, P.V.N., Choudhury, S.B. *et al.* (2018) Influence of

meteorological parameters on atmospheric CO2 at Bharati, the Indian Antarctic research station. Polar Research 37: 1442072.

227. Pelletier, E., Delille, D., and Delille, B. (2004) Crude oil bioremediation in sub-Antarctic intertidal sediments: chemistry and toxicity of oiled residues. Marine Environmental Research 57: 311-327.

228. Piñones, A., and Fedorov, A.V. (2016) Projected changes of Antarctic krill habitat by the end of the 21st century. Geophysical Research Letters 43: 85808589.

229. Poland, J.S., Riddle, M.J., and Zeeb, B.A. (2003) Contaminants in the Arctic and the Antarctic: a comparison of sources, impacts, and remediation options. The Polar Record 39: 369.

230. Potapowicz, J., Szumińska, D., Szopińska, M., and Polkowska, Ż. (2019) The influence of global climate change on the environmental fate of anthropogenic pollution released from the permafrost: part I. Case study of Antarctica. Science of the Total Environment 651: 1534-1548.

231. Prajith, R., Rout, R., Kumbhar, D., Mishra, R., Sahoo, B., and Sapra, B. (2019) Measurements of radon (222 Rn) and thoron (220 Rn) exhalations and their decay product concentrations at Indian Stations in Antarctica. Environmental Earth Sciences 78: 35.

232. Pratt, D.L. (1960) Mechanical equipment in Antarctica and the relation of laboratory research to field performance.

233. Ramesh, K., and Soni, V. (2018) Perspectives of Antarctic weather monitoring and research efforts. Polar Science 18: 183-188.

234. Ranjan, R.R., Ghadai, N.C., Iyer, K., Joshi, H., Dhar, A., Rawat, R., and Pathan, B. (2015) Study of Mass Distribution of Near-Surface Aerosols At Maitri, Antarctica.

235. Rautio, M., and Vincent, W.F. (2006) Benthic and pelagic food resources for zooplankton in shallow high-latitude lakes and ponds. Freshwater Biology 51: 1038-1052.

236. Raymond, T., King, C., Raymond, B., Stark, J., and Snape, I. (2017) Oil pollution in Antarctica. In: Gulf Professional Publishing Oxford, United Kingdom, pp. 759-803.

237. Reed, S., Clark, M., Thompson, R., and Hughes, K.A. (2018) Microplastics in marine sediments near Rothera research station, Antarctica. Marine pollution bulletin 133: 460-463.

238. Rich, R. (1982) A minerals regime for Antarctica. International and Comparative Law Quarterly: 709-725.

239. Richards, F. (2015) Report on Greatest Risk of Non-native Species entering Antarctica, Plans of Identification and Eradication for Postgraduate Certificate in Antarctic Studies.

240. Ridoux, V., and Offredo, C. (1989) The diets of five summer breeding seabirds in Adélie Land, Antarctica. Polar Biology 9: 137-145.

241. Riffenburgh, B. (2006) Encyclopedia of the Antarctic: Routledge.

242. Rintoul, S.R. (2010) Antarctic circumpolar current. Ocean currents: 196208.

243. Rintoul, S.R., Chown, S.L., DeConto, R.M., England, M.H., Fricker, H.A., Masson-Delmotte, V. *et al.* (2018) Choosing the future of Antarctica. Nature 558: 233-241.

244. Roseblade, R. (1996) Disposal of Garbage Aboard RAN Ships-A Review of Possible Options.

245. Ruberto, L., Vazquez, S.C., and Mac Cormack, W.P. (2003) Effectiveness of the natural bacterial flora, biostimulation and bioaugmentation on the bioremediation of a hydrocarbon-contaminated Antarctic soil. International Biodeterioration & Biodegradation 52: 115-125.

246. Sabaka, T.J., Tøffner-Clausen, L., Olsen, N., and Finlay, C.C. (2018) A comprehensive model of Earth's magnetic field determined from 4 years of Swarm satellite observations. Earth, Planets and Space 70: 1-26.

247. Sánchez, R.A., and McIvor, E. (2007) The Antarctic Committee for Environmental Protection: past, present, and future. The Polar Record 43: 239.

248. Sand, P.H. (2008) Japan's 'research whaling'in the Antarctic Southern Ocean and the North Pacific Ocean in the face of the

Endangered Species Convention (CITES). Review of European Community & International Environmental Law 17: 56-71.

249. Santiago, I.F., Alves, T.M., Rabello, A., Junior, P.A.S., Romanha, A.J., Zani, C.L. *et al.* (2012) Leishmanicidal and antitumoral activities of endophytic fungi associated with the Antarctic angiosperms Deschampsia Antarctica Desv. and Colobanthus quitensis (Kunth) Bartl. Extremophiles 16: 95-103.

250. Santos, I.R., Silva-Filho, E.V., Schaefer, C.E., Albuquerque-Filho, M.R., and Campos, L.S. (2005) Heavy metal contamination in coastal sediments and soils near the Brazilian Antarctic Station, King George Island. Marine pollution bulletin 50: 185-194.

251. Scambos, T.A., Campbell, G.G., Pope, A., Haran, T., Muto, A., Lazzara, M. *et al.* (2018) Ultralow Surface Temperatures in East Antarctica From Satellite Thermal Infrared Mapping: The Coldest Places on Earth. Geophysical Research Letters 45: 6124-6133.

52. Schopf, J., and Long, W. (1966) Coal metamorphism and igneous associations in Antarctica. In: ACS Publications.

253. Segev, A. (2002) Flood basalts, continental breakup and the dispersal of Gondwana: evidence for periodic migration of upwelling mantle flows (plumes). EGU Stephan Mueller Special Publication Series 2: 171-191.

254. Shannon, M. (2008) From zebra mussels to coqui frogs: public nuisance liability as a method to combat the introduction of invasive species. Environs: Envtl L & Pol'y J 32: 37.

255. Shepherd, A., Fricker, H.A., and Farrell, S.L. (2018) Trends and connections across the Antarctic cryosphere. Nature 558: 223-232.

256. Shi, G., Wang, X.-C., Li, Y., Trengove, R., Hu, Z., Mi, M. *et al.* (2019) Organic tracers from biomass burning in snow from the coast to the ice sheet summit of East Antarctica. Atmospheric Environment 201: 231-241.

257. Shivaji, S. (2015) Planococcus. Bergey's Manual of Systematics of Archaea and Bacteria: 1-11.

258. Shravage, B.V., Dayananda, K.M., Patole, M.S., and Shouche, Y.S. (2007) Molecular microbial diversity of a soil sample and detection of ammonia oxidizers from Cape Evans, Mcmurdo Dry Valley, Antarctica. Microbiological Research 162: 15-25.

259. Siingh, D., Singh, R., Gopalakrishnan, V., Selvaraj, C., and Panneerselvam, C. (2013) Fair-weather atmospheric electricity study at Maitri (Antarctica). Earth, Planets and Space 65: 1541-1553.

260. Silva, L.J., Crevelin, E.J., Souza, D.T., Lacerda-Júnior, G.V., de Oliveira, V.M., Ruiz, A.L.T.G. *et al.* (2020) Actinobacteria from Antarctica as a source for anticancer discovery. Scientific Reports 10: 1-15.

261. Simas, F.N., Schaefer, C.E.G., Melo, V.F., Guerra, M.B., Saunders, M., and Gilkes, R.J. (2006) Clay-sized minerals in permafrost-affected soils (Cryosols) from King George Island, Antarctica. Clays and clay minerals 54: 721-736.

262. Simões, F., Pfaff, R., Berthelier, J.-J., and Klenzing, J. (2012) A review of low-frequency electromagnetic wave phenomena related to troposphericionospheric coupling mechanisms. Space science reviews 168: 551-593.

263. Singh, D., and Semwal, R. (2006) Bryoflora of Schirmacher Oasis, East Antarctica: A preliminary study.

264. Singh, L., and Agarwal, M. (2006) Air Microflora of Antarctica.

265. Singh, S.M., and Nayaka, S. (2017) Contributions to the floral diversity of Schirmacher Oasis and Larsemann Hills, Antarctica. In Proc Indian Natn Sci Acad, pp. 469-481.

266. Sinha, A., Kumar, P., Rana, N.P., Islam, R., and Dwivedi, Y.K. (2019) Impact of internet of things (IoT) in disaster management: a task-technology fit perspective. Annals of Operations Research 283: 759-794.

267. Smetacek, V., and Nicol, S. (2005) Polar ocean ecosystems in a changing world. Nature 437: 362-368.

268. Smith, J.J., and Riddle, M. (2009) Sewage disposal and wildlife health in Antarctica. In Health of Antarctic Wildlife: Springer, pp. 271-315.

269. Smith, R.L. (1996) Introduced plants in Antarctica: potential impacts and conservation issues. Biological Conservation 76: 135-146.

270. Snape, I., Ferguson, S., and Revill, A. (2003) Constraints on rates of natural attenuation and in situ bioremediation of petroleum spills in Antarctica. Assessment and remediation of contaminated sites in Arctic and cold climates (Proceedings) St Joseph's Print Group Inc, Edmonton AB: 257-261.

271. Snape, I., Riddle, M.J., Stark, J.S., Cole, C.M., King, C.K., Duquesne, S., and Gore, D.B. (2001) Management and remediation of contaminated sites at Casey Station, Antarctica. Polar record 37: 199-214.

272. Soni, V., Sateesh, M., Das, A.K., and Peshin, S. (2017) Progress in meteorological studies around Indian stations in Antarctica. In Proc Indian Natl Sci Acad, pp. 461-467.

273. Splettstoesser, J. (2000) IAATO's stewardship of the Antarctic environment: A history of tour operator's concern for a vulnerable part of the world. International Journal of Tourism Research 2: 47-55.

274. Stark, J.S., Snape, I., and Riddle, M.J. (2006) Abandoned Antarctic waste disposal sites: monitoring remediation outcomes and limitations at Casey Station. In: Wiley Online Library.

275. Suaria, G., Perold, V., Lee, J.R., Lebouard, F., Aliani, S., and Ryan, P.G. (2020) Floating macro-and microplastics around the Southern Ocean: Results from the Antarctic Circumnavigation Expedition. Environment International 136: 105494.

276. Szopińska, M., Szumińska, D., Bialik, R.J., Dymerski, T., Rosenberg, E., and Polkowska, Ż. (2019) Determination of polycyclic aromatic hydrocarbons (PAHs) and other organic pollutants in freshwaters on the western shore of Admiralty Bay (King George Island, Maritime Antarctica). Environmental Science and Pollution Research 26: 18143-18161.

277. Szpalski, M. (2006) The World Factbook 2005 The World Factbook 2005, 2005. Spine 31: 1515-1519.

278. Talley, L.D. (2013) Closure of the global overturning circulation through the Indian, Pacific, and Southern Oceans: Schematics and transports. Oceanography 26: 80-97.

279. Tamm, S. (2018) Peace vs. compliance in Antarctica: inspections and the environment. The Polar Journal 8: 333-350.

280. Taori, A., Das, S., Goenka, R., Gharai, B., Rao, P., Seshasai, M., and Thakur, J. (2018) Round-the-clock measurements of aerosol optical thickness over Antarctica made using a Dual Imager System during January-February 2017. Remote Sensing Letters 9: 1089-1098.

281. Tarling, D. (1971) Gondwanaland, palaeomagnetism and continental drift. Nature 229: 17-21.

282. Tin, T., Sovacool, B.K., Blake, D., Magill, P., El Naggar, S., Lidstrom, S. *et al.* (2010) Energy efficiency and renewable energy under extreme conditions: Case studies from Antarctica. Renewable Energy 35: 1715-1723.

283. Tin, T., Fleming, Z.L., Hughes, K.A., Ainley, D.G., Convey, P., Moreno, C.A. *et al.* (2008) Impacts of local human activities on the Antarctic environment. Antarctic Science 21: 3-33.

284. Tin, T., Fleming, Z.L., Hughes, K.A., Ainley, D., Convey, P., Moreno, C. *et al.* (2009) Impacts of local human activities on the Antarctic environment. Antarctic Science 21: 3-33.

285. Tiwari, A., and Nayak, G. (2007) Water abundance and effect of glacier melting at Priyadarshini lake in Antarctica. Journal of Institution of Engineers India (Environmental Engineering) 88 27-32.

286. Tiwari, A., Pandey, P., and Khare, N. (2006a) Environmental Risk Assessment and Environmental Code of Conduct for Antarctic Protected Areas. http://hdl.handle.net/123456789/651.

287. Tiwari, A., da Lima Leitao, T.M., and Khare, N. (2020) Antarctic Environmental Studies over the Last 35 Years of the Indian Antarctic Expedition. In: Engineering and Communications in Antarctica: Springer, pp. 123-134.

288. Tiwari, A.K. (2006) Environmental Status at Indian Polar Research Station Maitri-A Comprehensive Study. http://npdc.ncpor.res.in/npdc/search_mf_data.action?

289. Tiwari, A.K. (2008) Environmental Awareness in Indian Antarctic Scientific Programme. Journal of Institution of Engineers India (Environmental Engineering) 88: 21-24.

290. Tiwari, A.K. (2017) Environmental Monitoring around Indian Stations. Proceedings of the Indian National Science Academy 90 (2). 10.16943/ptinsa/2017/48964.

291. Tiwari, A.K., and Kulkarni, S. (2006) Environmental Status at Indian Polar Research Station Maitri-A Comprehensive Study. http://14.139.119.23:8080/dspace/bitstream/123456789/581/3/ARTICLE+15.pdf.

292. Tiwari, A.K., Kulkarni, S., Ramteke, D., and Nayak, G. (2006b) Environment quality at Maitri station in Antarctica. Indian Journal of Environmental Health 48(3):191-198.

293. Tiwari, V., Singh, B., Rao, M.V., and Mishra, D. (2006c) Absolute gravity measurements in India and Antarctica. Current Science: 686-689.

294. Tosi, S., Kostadinova, N., Krumova, E., Pashova, S., Dishliiska, V., Spassova, B. *et al.* (2010) Antioxidant enzyme activity of filamentous fungi isolated from Livingston Island, Maritime Antarctica. Polar Biology 33: 1227-1237.

295. Treaty, A. (1959) Antarctic Treaty: Washington DC, USA.

296. Turner, J., Anderson, P., Lachlan-Cope, T., Colwell, S., Phillips, T., Kirchgaessner, A. *et al.* (2009a) Record low surface air temperature at Vostok station, Antarctica. Journal of Geophysical Research: Atmospheres 114.

297. Turner, J., Bindschadler, R., Convey, P., Di Prisco, G., Fahrbach, E., Gutt, J., Hodgson, D.A., Mayewski, P.A., and Summerhayes, C.P. (2009a) Antarctic climate change and the environment. 526 pp., 2009. Cambridge, SCAR. ISBN 978 0 948277 22 1.

298. Turner, J., Colwell, S.R., Marshall, G.J., Lachlan-Cope, T.A., Carleton, A.M., Jones, P.D. *et al.* (2005b) Antarctic climate

change during the last 50 years. International journal of Climatology 25: 279-294.

299. Van Den Broeke, M. (2004) On the role of Antarctica as heat sink for the global atmosphere. In Journal de Physique IV (Proceedings): EDP sciences, pp. 115-124.

300. Van den Broeke, M., and Van Lipzig, N. (2003) Factors controlling the nearsurface wind field in Antarctica. Monthly Weather Review 131: 733-743.

301. VanderZwaag, D.L. (2010) 3. Law of the Sea and Governance of Shipping in the Arctic and Antarctic. Polar Law Textbook: 45.

302. Vecchiato, M., Zambon, S., Argiriadis, E., Barbante, C., Gambaro, A., and Piazza, R. (2015) Polychlorinated biphenyls (PCBs) and polybrominated diphenyl ethers (PBDEs) in Antarctic ice-free areas: influence of local sources on lakes and soils. Microchemical Journal 120: 26-33.

303. Venkatramanan, S., Wu, S., Shi, B., Marathe, A., Marathe, M., Eubank, S. *et al.* (2017) Towards robust models of food flows and their role in invasive species spread. In: 2017 IEEE International Conference on Big Data (Big Data): IEEE, pp. 435-444.

304. Vidas, D. (2002) The Protocol on Environmental Protection to the Antarctic Treaty: A Ten-Year Review. Yearbook of International Co-operation on Environment and Development 2003: 51-60.

305. Vodopivez, C., Curtosi, A., Villaamil, E., Smichowski, P., Pelletier, E., and Mac Cormack, W.P. (2015) Heavy metals in sediments and soft tissues of the Antarctic clam Laternula elliptica: More evidence as a? possible biomonitor of coastal marine pollution at high latitudes? Science of the Total Environment 502: 375-384.

306. Waller, C.L., Griffiths, H.J., Waluda, C.M., Thorpe, S.E., Loaiza, I., Moreno, B. *et al.* (2017) Microplastics in the Antarctic marine system: an emerging area of research. Science of the Total Environment 598: 220-227.

307. Walworth, J., Pond, A., Snape, I., Rayner, J., Ferguson, S., and Harvey, P. (2007) Nitrogen requirements for maximizing petroleum bioremediation in a sub-Antarctic soil. Cold Regions Science and Technology 48: 84-91.

308. Wasson, J.T. (1990) Ungrouped iron meteorites in Antarctica: Origin of anomalously high abundance. Science 249: 900-902.

309. Weller, R., Minikin, A., Petzold, A., Wagenbach, D., and König-Langlo, G. (2013) Characterization of long-term and seasonal variations of black carbon (BC) concentrations at Neumayer, Antarctica. Atmospheric Chemistry and Physics 13: 1579-1590.

310. Wery, N., Gerike, U., Sharman, A., Chaudhuri, J.B., Hough, D.W., and Danson, M.J. (2003) Use of a packed-column bioreactor for isolation of diverse protease-producing bacteria from antarctic soil. Applied and environmental microbiology 69: 1457-1464.

311. Wharton, D. (2003) The environmental physiology of Antarctic terrestrial nematodes: a review. Journal of Comparative Physiology B 173: 621-628.

312. Williams, G.D., Nicol, S., Aoki, S., Meijers, A.J., Bindoff, N.L., Iijima, Y. *et al.* (2010) Surface oceanography of BROKE-West, along the Antarctic margin of the south-west Indian Ocean (30–80 ° E). Deep-Sea Research Part II: Topical Studies in Oceanography 57: 738-757.

313. Winkelmann, R., Levermann, A., Martin, M.A., and Frieler, K. (2012) Increased future ice discharge from Antarctica owing to higher snowfall. Nature 492: 239-242.

314. Yi, H., Oh, H.-M., Lee, J.-H., Kim, S.-J., and Chun, J. (2005) *Flavobacterium antarcticum* sp. nov., a novel psychrotolerant bacterium isolated from the Antarctic. International journal of systematic and evolutionary microbiology 55: 637-641.

315. Zangrando, R., Barbaro, E., Vecchiato, M., Kehrwald, N.M., Barbante, C., and Gambaro, A. (2016) Levoglucosan and phenols in Antarctic marine, coastal and plateau aerosols. Science of the Total Environment 544: 606-616.

316. Zerbini, A.N., Adams, G., Best, J., Clapham, P.J., Jackson, J.A., and Punt, A.E. (2019) Assessing the recovery of an Antarctic predator from historical exploitation. Royal Society open science 6: 190368.

317. Zhang, N., and Talalay, P.G.e. (2021) Antarctic oil and mineral resources: a subject off limits or future reality? Earth and Space Science Open Archive ESSOAr.

SOURCE OF FIGURES

[1] Source: https://www.bing.com/images

[2] Source: Photographed by Anoop Tiwari (Personal communication)

[3] Source: Photographed by Anoop Tiwari (Personal communication)

[4] Source: Google Earth

[5] Source: NCPOR, Polar Photo Repository

[6] Source: NCPOR, Polar Photo Repository

[7] Source: Photographed by Anoop Tiwari (Personal communication)

[8] Source: Photographed by K. R. Das (Personal communication)

[9] Source: Photographed by Anoop Tiwari (Personal communication)

[10] Source: Photographed by Anoop Tiwari (Personal communication)

[11] Source: Photographed by Ajay Dhar (Personal communication)

[12] Source: Photographed by Anoop Tiwari (Personal communication)

[13] Source: Photographed by Anoop Tiwari (Personal communication)

[14] Source: Photographed by Anoop Tiwari (Personal communication)

[15] Source: Photographed by Anoop Tiwari (Personal communication)

[16] Source: Photographed by Anoop Tiwari (Personal communication)

[17] Source: http://www.coolantarctica.com

[18] Source: https://commons.wikimedia.org/wiki/File:BlueWhaleWithCalf.jpg

[19] Source: https://commons.wikimedia.org/wiki/File:Antarctic_Fur_Seal_swimming_(6018836251).jpg

[20] Source: https://commons.wikimedia.org/wiki/File:South_pole_skiers.jpg

[21] Source: Photographed by Anoop Tiwari (Personal communication)

[22] Source: Photographed by Anoop Tiwari (Personal communication)

[23] Source: Photographed by Anoop Tiwari (Personal communication)

[24] Source: Photographed by Anoop Tiwari (Personal communication)

[25] Source: Photographed by Anoop Tiwari (Personal communication)

[26] Source: Photographed by Anoop Tiwari (Personal communication)

[27] Source: Photographed by Anoop Tiwari (Personal communication)

[28] Source: Photographed by Anoop Tiwari (Personal communication)

[29] Source: Photographed by Anoop Tiwari (Personal communication)

[30] Source: Photographed by Anoop Tiwari (Personal communication)

ANNEXURE

The Protocol on Environmental Protection to the Antarctic Treaty

Preamble

The States Parties to this Protocol to the Antarctic Treaty hereinafter referred to as the Parties,

Convinced of the need to enhance the protection of the Antarctic environment and dependent and associated ecosystems;

Convinced of the need to strengthen the Antarctic Treaty system so as to ensure that Antarctica shall continue forever to be used exclusively for peaceful purposes and shall not become the scene or object of international discord;

Bearing in mind the special legal and political status of Antarctica and the special responsibility of the Antarctic Treaty Consultative Parties to ensure that all activities in Antarctica are consistent with the purposes and principles of the Antarctic Treaty;

Recalling the designation of Antarctica as a Special Conservation Area and other measures adopted under the Antarctic Treaty system to protect the Antarctic environment and dependent and associated ecosystems;

Acknowledging further the unique opportunities Antarctica offers for scientific monitoring of and research on processes of global as well as regional importance;

Reaffirming the conservation principles of the Convention on the Conservation of Antarctic Marine Living Resources;

Convinced that the development of a comprehensive regime for the protection of the Antarctic environment and dependent and associated ecosystems is in the interest of mankind as a whole;

Desiring to supplement the Antarctic Treaty to this end;

Have agreed as follows:

Article 1

Definitions

For the purposes of this Protocol:

(a) "The Antarctic Treaty" means the Antarctic Treaty done at Washington on 1st December 1959;

(b) "Antarctic Treaty area" means the area to which the provisions of the Antarctic Treaty apply in accordance with Article VI of that Treaty;

(c) "Antarctic Treaty Consultative Meetings" means the meetings referred to in Article IX of the Antarctic Treaty;

(d) "Antarctic Treaty Consultative Parties" means the Contracting Parties to the Antarctic Treaty entitled to appoint representatives to participate in the meetings referred to in Article IX of that Treaty;

(e) "Antarctic Treaty system" means the Antarctic Treaty, the measures in effect under that Treaty, its associated separate international instruments in force and the measures in effect under those instruments;

(f) "Arbitral Tribunal" means the Arbitral Tribunal established in accordance with the Schedule to this Protocol, which forms an integral part thereof;

(g) "Committee" means the Committee for Environmental Protection established in accordance with Article 11.

Article 2

Objective and Designation

The Parties commit themselves to the comprehensive protection of the Antarctic environment and dependent and associated ecosystems and hereby designate Antarctica as a natural reserve, devoted to peace and science.

Article 3

Environmental Principles:

1 The protection of the Antarctic environment and dependent and associated ecosystems and the intrinsic value of Antarctica, including its wilderness and aesthetic values and its value as an area for the conduct of scientific research, in particular research essential to understanding the global environment, shall be fundamental considerations in the planning and conduct of all activities in the Antarctic Treaty area.

2 To this end:

(a) activities in the Antarctic Treaty area shall be planned and conducted so as to limit adverse impacts on the Antarctic environment and dependent and associated ecosystems;

(b) activities in the Antarctic Treaty area shall be planned and conducted so as to avoid:

(i) adverse effects on climate or weather patterns;

(ii) significant adverse effects on air or water quality;

(iii) significant changes in the atmospheric, terrestrial (including aquatic), glacial or marine environments;

(iv) detrimental changes in the distribution, abundance or productivity of species or populations of species of fauna and flora;

(v) further jeopardy to endangered or threatened species or populations of such species; or

(vi) degradation of, or substantial risk to, areas of biological, scientific, historic, aesthetic or wilderness significance;

(c) activities in the Antarctic Treaty area shall be planned and conducted on the basis of information sufficient to allow prior assessments of, and informed judgements about, their possible impacts on the Antarctic environment and dependent and associated ecosystems and on the value of Antarctica for the conduct of scientific research; such judgements shall take account of:

(i) the scope of the activity, including its area, duration and intensity;

(ii) the cumulative impacts of the activity, both by itself and in combination with other activities in the Antarctic Treaty area;

(iii) whether the activity will detrimentally affect any other activity in the Antarctic Treaty area;

(iv) whether technology and procedures are available to provide for environmentally safe operations;

(v) whether there exists the capacity to monitor key environmental parameters and ecosystem components so as to identify and provide early warning of any adverse effects of the activity and to provide for such modification of operating procedures as may be necessary in the light of the results of monitoring or increased knowledge of the Antarctic environment and dependent and associated ecosystems; and

(vi) whether there exists the capacity to respond promptly and effectively to accidents, particularly those with potential environmental effects;

(d) regular and effective monitoring shall take place to all assessments of the impacts of ongoing activities, including the verification of predicted impacts;

(e) regular and effective monitoring shall take place to facilitate early detection of the possible unforeseen effects of activities carried on both within and outside the Antarctic Treaty area on the Antarctic environment and dependent and associated ecosystems.

3 Activities shall be planned and conducted in the Antarctic Treaty area so as to accord priority to scientific research and to preserve the value of Antarctica as an area for the conduct of such research, including research essential to understanding the global environment.

4 Activities undertaken in the Antarctic Treaty area pursuant to scientific research programs, tourism and all other governmental and non-governmental activities in the Antarctic Treaty area for which advance notice is required in accordance with Article VII (5) of the Antarctic Treaty, including associated logistic activities, shall:

(a) take place in a manner consistent with the principles in this Article; and

(b) be modified, suspended or cancelled if they result in or threaten to result in impacts upon the Antarctic environment or dependent or associated ecosystems inconsistent with those principles.

Article 4

Relationship with other Components of the Antarctic Treaty System

1 This Protocol shall supplement the Antarctic Treaty and shall neither modify nor amend that Treaty.

2 Nothing in this Protocol shall derogate from the rights and obligations of the Parties to this Protocol under the other international instruments in force within the Antarctic Treaty system.

Article 5

Consistency with other Components of the Antarctic Treaty System

The Parties shall consult and co-operate with the Contracting Parties to the other international instruments in force within the Antarctic Treaty system and their respective institutions with a view to ensuring the achievement of the objectives and principles of this Protocol and avoiding any interference with the achievement of the objectives and principles of those instruments or any inconsistency between the implementation of those instruments and this Protocol.

Article 6

Co-operation

1 The Parties shall co-operate in the planning and conduct of activities in the Antarctic Treaty area. To this end, each Party shall endeavour to:

(a) promote co-operative programs of scientific, technical and educational value, concerning the protection of the Antarctic environment and dependent and associated ecosystems;

(b) provide appropriate assistance to other Parties in the preparation of environmental impact assessments;

(c) provide to other Parties upon request information relevant to any potential environmental risk and assistance to minimize the effects of accidents which may damage the Antarctic environment or dependent and associated ecosystems;

(d) consult with other Parties with regard to the choice of sites for prospective stations and other facilities so as to avoid the cumulative impacts caused by their excessive concentration in any location;

(e) where appropriate, undertake joint expeditions and share the use of stations and other facilities; and

(f) carry out such steps as may be agreed upon at Antarctic Treaty Consultative Meetings.

2 Each Party undertakes, to the extent possible, to share information that may be helpful to other Parties in planning and conducting their activities in the Antarctic Treaty area, with a view to the protection of the Antarctic environment and dependent and associated ecosystems.

3 The Parties shall co-operate with those Parties which may exercise jurisdiction in areas adjacent to the Antarctic Treaty area with a view to ensuring that activities in the Antarctic Treaty area do not have adverse environmental impacts on those areas.

Article 7

Prohibition of Mineral Resource Activities

Any activity relating to mineral resources, other than scientific research, shall be prohibited.

Article 8

Environmental Impact and Assessment

1 Proposed activities referred to in paragraph 2 below shall be subject to the procedures set out in Annex I for prior assessment of the impacts of those activities on the Antarctic environment or on dependent or associated ecosystems according to whether those activities are identified as having:

(a) less than a minor or transitory impact;

(b) a minor or transitory impact; or

(c) more than a minor or transitory impact.

2 Each Party shall ensure that the assessment procedures set out in Annex I are applied in the planning processes leading to decisions about any activities undertaken in the Antarctic Treaty area pursuant to scientific research programs, tourism and all other governmental and non-governmental activities in the Antarctic Treaty area for which advance notice is required under Article VII (5) of the Antarctic Treaty, including associated logistic support activities.

3 The assessment procedures set out in Annex I shall apply to any change in an activity whether the change arises from an increase or decrease in the intensity of an existing activity, from the addition of activity, the decommissioning of a facility, or otherwise.

4 Where activities are planned jointly by more than one Party, the Parties involved shall nominate one of their number to co-ordinate the implementation of the environmental impact assessment procedures set out in Annex I.

Article 9

Annexs

1. The annexes to this Protocol shall form an integral part thereof.

2 Annexs, additional to Annexs I-IV, may be adopted and become effective in accordance with Article IX of the Antarctic Treaty.

3 Amendments and modifications to Annexs may be adopted and become effective in accordance with Article IX of the Antarctic Treaty, provided that any Annex may itself make provision for amendments and modifications to become effective on an accelerated basis.

4 Annexs and any amendments and modifications thereto which have become effective in accordance with paragraphs 2 and 3 above shall, unless an Annex itself provides otherwise in respect of the entry into effect of any amendment or modification thereto, become effective for a Contracting Party to the Antarctic Treaty which is not an Antarctic Treaty Consultative Party, or which was not an Antarctic Treaty Consultative Party at the time of the adoption when notice of approval of that Contracting Party has been received by the Depositary.

5 Annexs shall, except to the extent that an Annex provides otherwise, be subject to the procedures for dispute settlement set out in Articles 18 to 20.

Article 10

Antarctic Treaty Consultative Meetings

1 Antarctic Treaty Consultative Meetings shall, drawing upon the best scientific and technical advice available:

(a) define, in accordance with the provisions of this Protocol, the general policy for the comprehensive protection of the Antarctic environment and dependent and associated ecosystems; and

(b) adopt measures under Article IX of the Antarctic Treaty for the implementation of this Protocol.

2 Antarctic Treaty Consultative Meetings shall review the work of the Committee and shall draw fully upon its advice and recommendations in carrying out the tasks referred to in paragraph 1 above, as well as upon the advice of the Scientific Committee on Antarctic Research.

Article 11

Committee for Environmental Protection

1 There is hereby established the Committee for Environmental Protection.

2 Each Party shall be entitled to be a member of the Committee and to appoint a representative who may be accompanied by experts and advisers.

3 Observer status in the Committee shall be open to any Contracting Party to the Antarctic Treaty which is not a Party to this Protocol.

4 The Committee shall invite the President of the Scientific Committee on Antarctic Research and the Chairman of the Scientific Committee for the Conservation of Antarctic Marine Living Resources to participate as observers at its sessions. The Committee may also, with the approval of the Antarctic Treaty Consultative Meeting, invite such other relevant scientific, environmental and technical organisations which can contribute to its work to participate as observers at its sessions.

5 The Committee shall present a report on each of its sessions to the Antarctic Treaty Consultative Meeting. The report shall cover all matters considered at the session and shall reflect the views expressed. The report shall be circulated to the Parties and observers attending the session, and shall thereupon be made publicly available.

6 The Committee shall adopt its rules of procedure which shall be subject to approval by the Antarctic Treaty Consultative Meeting.

Article 12

Functions of the Committee

1 The functions of the Committee shall be to provide advice and formulate recommendations to the Parties in connection with the implementation of this Protocol, including the operation of its Annexs, for consideration at Antarctic Treaty Consultative Meetings, and to perform such other functions as may be referred to it by the Antarctic Treaty Consultative Meetings. In particular, it shall provide advice on:

(a) the effectiveness of measures taken pursuant to this Protocol;

(b) the need to update, strengthen or otherwise improve such measures;

(c) the need for additional measures, including the need for additional Annexs, where appropriate;

(d) the application and implementation of the environmental impact assessment procedures set out in Article 8 and Annex I;

(e) means of minimizing or mitigating environmental impacts of activities in the Antarctic Treaty area;

(f) procedures for situations requiring urgent action, including response action in environmental emergencies;

(g) the operation and further elaboration of the Antarctic Protected Area system;

(h) inspection procedures, including formats for inspection reports and checklists for the conduct of inspections;

(i) the collection, archiving, exchange and evaluation of information related to environmental protection;

(j) the state of the Antarctic environment; and

(k) the need for scientific research, including environmental monitoring, related to the implementation of this Protocol.

2 In carrying out its functions, the Committee shall, as appropriate, consult with the Scientific Committee on Antarctic Research, the Scientific Committee for the Conservation of Antarctic Marine Living Resources and other relevant scientific, environmental and technical organisations.

Article 13

Compliance with this Protocol

1 Each Party shall take appropriate measures within its competence, including the adoption of laws and regulations, administrative actions and enforcement measures, to ensure compliance with this Protocol.

2 Each Party shall exert appropriate efforts, consistent with the Charter of the United Nations, to the end that no one engages in any activity contrary to this Protocol.

3 Each Party shall notify all other Parties of the measures it takes pursuant to paragraphs 1 and 2 above.

4 Each Party shall draw the attention of all other Parties to any activity which in its opinion affects the implementation of the objectives and principles of this Protocol.

5 The Antarctic Treaty Consultative Meetings shall draw the attention of any State which is not a Party to this Protocol to any activity undertaken by that State, its agencies, instrumentalities, natural or juridical persons, ships, aircraft or other means of transport which affects the implementation of the objectives and principles of this Protocol.

Article 14

Inspection

1 In order to promote the protection of the Antarctic environment and its dependent and associated ecosystems, and to ensure compliance with this Protocol, the Antarctic Treaty Consultative Parties shall arrange, individually or collectively, for inspections by observers to be made in accordance with Article VII of the Antarctic Treaty.

2 Observers are:

(a) observers designated by any Antarctic Treaty Consultative Party who shall be nationals of that Party, and

(b) any observers designated at Antarctic Treaty Consultative Meetings to carry out inspections under procedures to be established by an Antarctic Treaty Consultative Meeting.

3 Parties shall co-operate fully with observers undertaking inspections, and shall ensure that during inspections, observers are given access to all parts of stations, installations, equipment, ships and aircraft open to inspection under Article VII (3) of the Antarctic Treaty, as well as to all records maintained thereon which are called for pursuant to this Protocol.

4 Reports of inspections shall be sent to the Parties whose stations, installations, equipment, ships or aircraft are covered by the reports. After those Parties have been given the opportunity to comment, the reports and any comments thereon shall be circulated to all the Parties and the Committee, considered at Antarctic Treaty Consultative Meeting, and thereafter made publicly available.

Article 15

Emergency Response Action

1 In order to respond to environmental emergencies in ^ Antarctic Treaty area, each Party agrees to:

(a) provide for prompt and effective response action to such emergencies which might arise in the performance of scientific research programs, tourism and all other governmental and non-governmental activities in the Antarctic Treaty area for which advance notice is required under Article VII (5) of the Antarctic Treaty, including associated logistic support activities; and

(b) establish contingency plans for response to incidents with potentially adverse effects on the Antarctic environment or dependent and associated ecosystems.

2 To this end, the Parties shall:

(a) cooperate in the formulation and implementation of such contingency plans; and

(b) establish procedures for immediate notification of, and co-operative response to, environmental emergencies.

3 In the implementation of this Article, the Parties shall draw upon the advice of the appropriate international organisations.

Article 16

Liability

Consistent with the objectives of this Protocol for the comprehensive protection of the Antarctic environment and dependent and associated ecosystems, the Parties undertake to elaborate rules and procedures relating to liability for damage arising from activities taking place in the Antarctic Treaty area and covered by this Protocol. Those rules and procedures shall be included in one or more Annexs to be adopted in accordance with Article 9 (2).

Article 17

Annual Report by Parties

1 Each Party shall report annually on the steps taken to implement this Protocol. Such reports shall include notifications made in accordance with Article 13 (3), contingency plans established in accordance with Article 15 and any other notifications and information called for pursuant to this Protocol for which there is no other provision concerning the circulation and exchange of information.

2 Reports made in accordance with paragraph 1 above shall be circulated to all Parties and the Committee, considered at the next Antarctic Treaty Consultative Meeting, and made publicly available.

Article 18

Dispute Settlement

If a dispute arises concerning the interpretation or application of this Protocol, the parties to the dispute shall, at the request of any one of them, consult among themselves as soon as possible with a view to having the dispute resolved by negotiation, inquiry, mediation, conciliation, arbitration, judicial settlement or other agreeable to all concerned parties.

Article 19

Choice of Dispute Settlement Procedure

1 Each Party, when signing, ratifying, accepting, approving or acceding to this Protocol, or at any time thereafter, may choose, by written declaration, one or both of the following means for the settlement of disputes concerning the interpretation or application of Articles 7, 8 and 15 and, except to the extent that an Annex provides otherwise, the provisions of any Annex and, insofar as it relates to these Articles and provisions, Article 13:

(a) the International Court of Justice;

(b) the Arbitral Tribunal.

2 A declaration made under paragraph 1 above shall not affect the operation of Article 18 and Article 20 (2).

3 A Party which has not made a declaration under paragraph 1 above or in respect of which a declaration is no longer in force shall be deemed to have accepted the competence of the Arbitral Tribunal.

4 If the parties to a dispute have accepted the same means for the settlement of a dispute, the dispute may be submitted only to that procedure, unless the parties otherwise agree.

5 If the parties to a dispute have not accepted the same means for the settlement of a dispute, or if they have both accepted both means, the dispute may be submitted only to the Arbitral Tribunal, unless the parties otherwise agree.

6 A declaration made under paragraph 1 above shall remain in force until it expires in accordance with its terms or until three months after written notice of revocation has been deposited with the Depositary.

7 A new declaration, a notice of revocation or the expiry of a declaration shall not in any way affect proceedings pending before the International Court of Justice or the Arbitral Tribunal unless the parties to the dispute otherwise agree.

8 Declarations and notices referred to in this Article shall be deposited with the Depositary who shall transmit copies thereof to all Parties.

Article 20

Dispute Settling Procedure

1 If the parties to a dispute concerning the interpretation or application of Articles 7, 8 or 15 or, except to the extent that an Annex provides otherwise, the provisions of any Annex or, insofar as it relates to these Articles and provisions, Article 13, have not agreed on a means for resolving it within 12 months of the request for consultation pursuant to Article 18, the dispute shall be referred, at the request of any party to the dispute, for settlement in accordance with the procedure determined by Article 19 and (5).

2 The Arbitral Tribunal shall not be competent to decide or rule upon any matter within the scope of Article IV of the Antarctic

Treaty. In addition, nothing in this Protocol shall be interpreted as conferring competence or jurisdiction on the International Court of Justice or any other tribunal established for the purpose of settling disputes between Parties to decide or otherwise rule upon any matter within the scope of Article IV of the Antarctic Treaty.

Article 21

Signature

This Protocol shall be open for signature at Madrid on the 4th of October 1991 and thereafter at Washington until the 3rd of October 1992 by any State which is a Contracting Party to the Antarctic Treaty.

Article 22

Ratification, Acceptance, Approval or Accession

1 This Protocol is subject to ratification, acceptance or approval by signatory States.

2 After the 3rd of October 1992 this Protocol shall be open for accession by any State which is a Contracting Party to the Antarctic Treaty.

3 Instruments of ratification, acceptance, approval or accession shall be deposited with the Government of the United States of America, hereby designated as the Depository.

4 After the date on which this Protocol has entered into force, the Antarctic Treaty Consultative Parties shall not act upon a notification regarding the entitlement of a Contracting Party to the Antarctic Treaty to appoint representatives to participate in Antarctic Treaty Consultative Meetings in accordance with Article IX (2) of the Antarctic Treaty unless that Contracting Party has first ratified, accepted, approved or acceded to this Protocol.

Article 23

Entry into Force

1 This Protocol shall enter into force on the thirtieth day following the date of deposit of instruments of ratification, acceptance,

approval or accession by all States which are Antarctic Treaty Consultative Parties at the date on which this Protocol is adopted.

2 For each Contracting Party to the Antarctic Treaty which, subsequent to the date of entry into force of this Protocol, deposits an instrument of ratification, acceptance, approval or accession, this Protocol shall enter into force on the thirtieth day following such deposit.

Article 24

Reservations

Reservations to this Protocol shall not be permitted.

Article 25

Modification or Amendment

1 Without prejudice to the provisions of Article 9, this Protocol may be modified or amended at any time in accordance with the procedures set forth in Article XII (1) (a) and (b) of the Antarctic Treaty.

2 If, after the expiration of 50 years from the date of entry into force of this Protocol, any of the Antarctic Treaty Consultative Parties so requests by a communication addressed to the Depositary, a conference shall be held as soon as practicable to review the operation of this Protocol.

3 A modification or amendment proposed at any Review Conference called pursuant to paragraph 2 above shall be adopted by a majority of the Parties, including three-quarters of the States which are Antarctic Treaty Consultative Parties at the time of adoption of this Protocol.

4 A modification or amendment adopted pursuant to paragraph 3 above shall enter into force upon ratification, acceptance, approval or accession by threequarters of the Antarctic Treaty Consultative Parties, including ratification, acceptance, approval or accession by all States which are Antarctic Treaty Consultative Parties at the time of adoption of this Protocol.